T0348664

Initiation to
Global Finslerian Geometry

North-Holland Mathematical Library

Board of Honorary Editors:

M. Artin, H.Bass, J. Eells, W. Feit, P.J. Freyd, F.W. Gehring, H. Halberstam, L.V. Hörmander, J.H.B. Kemperman, W.A.J. Luxemburg, F. Peterson, I.M. Singer and A.C. Zaanen

Board of Advisory Editors:

A. Björner, R.H. Dijkgraaf, A. Dimca, A.S. Dow, J.J. Duistermaat, E. Looijenga, J.P. May, I. Moerdijk, S.M. Mori, J.P. Palis, A. Schrijver, J. Sjöstrand, J.H.M. Steenbrink, F. Takens and J. van Mill

VOLUME 68

ELSEVIER

Amsterdam – Boston – Heidelberg – London – New York – Oxford
Paris – San Diego – San Francisco – Singapore – Sydney – Tokyo

Initiation to Global Finslerian Geometry

H. Akbar-Zadeh
Director of Research at C.N.R.S.
Paris
France

ELSEVIER

Amsterdam – Boston – Heidelberg – London – New York – Oxford
Paris – San Diego – San Francisco – Singapore – Sydney – Tokyo

ELSEVIER B.V.
Radarweg 29
P.O. Box 211, 1000 AE Amsterdam
The Netherlands

ELSEVIER Inc.
525 B Street, Suite 1900
San Diego, CA 92101-4495
USA

ELSEVIER Ltd
The Boulevard, Langford Lane
Kidlington, Oxford OX5 1GB
UK

ELSEVIER Ltd
84 Theobalds Road
London WC1X 8RR
UK

© 2006 Elsevier B.V. All rights reserved.

This work is protected under copyright by Elsevier B.V., and the following terms and conditions apply to its use:

Photocopying
Single photocopies of single chapters may be made for personal use as allowed by national copyright laws. Permission of the Publisher and payment of a fee is required for all other photocopying, including multiple or systematic copying, copying for advertising or promotional purposes, resale, and all forms of document delivery. Special rates are available for educational institutions that wish to make photocopies for non-profit educational classroom use.

Permissions may be sought directly from Elsevier's Rights Department in Oxford, UK: phone (+44) 1865 843830, fax (+44) 1865 853333, e-mail: permissions@elsevier.com. Requests may also be completed on-line via the Elsevier homepage (http://www.elsevier.com/locate/permissions).

In the USA, users may clear permissions and make payments through the Copyright Clearance Center, Inc., 222 Rosewood Drive, Danvers, MA 01923, USA; phone: (+1) (978) 7508400, fax: (+1) (978) 7504744, and in the UK through the Copyright Licensing Agency Rapid Clearance Service (CLARCS), 90 Tottenham Court Road, London W1P 0LP, UK; phone: (+44) 20 7631 5555; fax: (+44) 20 7631 5500. Other countries may have a local reprographic rights agency for payments.

Derivative Works
Tables of contents may be reproduced for internal circulation, but permission of the Publisher is required for external resale or distribution of such material. Permission of the Publisher is required for all other derivative works, including compilations and translations.

Electronic Storage or Usage
Permission of the Publisher is required to store or use electronically any material contained in this work, including any chapter or part of a chapter.

Except as outlined above, no part of this work may be reproduced, stored in a retrieval system or transmitted in any form or by any means, electronic, mechanical, photocopying, recording or otherwise, without prior written permission of the Publisher.
Address permissions requests to: Elsevier's Rights Department, at the fax and e-mail addresses noted above.

Notice
No responsibility is assumed by the Publisher for any injury and/or damage to persons or property as a matter of products liability, negligence or otherwise, or from any use or operation of any methods, products, instructions or ideas contained in the material herein. Because of rapid advances in the medical sciences, in particular, independent verification of diagnoses and drug dosages should be made.

First edition 2006

Library of Congress Cataloging in Publication Data
A catalog record is available from the Library of Congress.

British Library Cataloguing in Publication Data
A catalogue record is available from the British Library.

ISBN-13: 978-0-444-52106-4
ISBN-10: 0-444-52106-2
Series ISSN: 0924-6509

♾ The paper used in this publication meets the requirements of ANSI/NISO Z39.48-1992 (Permanence of Paper).
Printed in The Netherlands.

PREFACE

This book is an initiation to global methods in the study of the differential geometry of Finsler manifolds. It contains my research on the subject during the last twenty years. Most of it has been published in articles in different journals. I have brought it all together in a streamlined manner to offer a coherent vision to global differential Finslerian geometry.

The first three chapters form the foundation of Finslerian geometry. They contain the basic notions of global Finslerian geometry and lay the groundwork for the rest of the book. The treatment is deliberately kept transparent and simple so as to highlight the differences between Riemannian geometry and Finslerian geometry.

At the start it is best to note that Finslerian geometry is the most natural generalization of Riemannian geometry. Consequently the results established in the book are the generalizations of the Riemannian case. Our particular interest bears on complete or compact manifolds. The book contains detailed proofs of a certain number results that I have published in the Comptes Rendus de l'Académie des Sciences de Paris.

The book has eight chapters. Each chapter begins with a résumé of the results contained in it. The number of publications on Finslerian geometry is high. I indicate only a few of them in the bibliography.

I want to thank here my friend Dr. Cyrille de Souza who has helped me in the preparation of this book. My sincere thanks go to Mr. Sevenster, editor of Elsevier, for his kindness, patience and advice during the preparation of the manuscript. I am also thankful to Ms. Andy Deelen, Administrative Editor, for her suggestions for the final preparation of the text. Finally, I am grateful to Elsevier for its interest in the progress of this branch of differential geometry.

Introduction

Finslerian geometry is the most natural generalization of Riemannian geometry. In his dissertation (1854[1]) Riemann already imagined a generalization of his metric. Then P.Finsler in his thesis (1918[17]) generalized a certain number of theorems of classical differential geometry. Berwald contributed to the progress of this geometry [1928 [8]). But the connection introduced by him is not Euclidean and deprives the Finslerian geometry the simplicity and elegance of Riemannian geometry as Elie Cartan remarks in his book (1933, [13]). Unfortunately some scholars of Finslerain geometry have not paid heed to Cartan's observation.

In his work Cartan studied the geometry of Finslerian manifolds in the framework of metric manifolds with the help of a Euclidean connection. He introduced the notion of the manifold of line elements ([13],[14]) that is formed by a set of points and the direction starting from those points. The parallel transport defined by him preserves the length of vectors. With the introduction of Euclidean connection in the neighbourhood of each linear element, the manifold enjoys all the properties of a Euclidean manifold. In other words the manifold is locally Euclidean.

The different infinitesimal connections introduced by Cartan (linear, projective and conformal) can be dealt with in the same geometric framework: C. Ehresmann published an article on this topic with the title "infinitesimal connections in a differentiable fibre bundle" (1950, pp. 29-55 [16]) in the context of any Lie group provided a general framework for the introduction of connections. A clear treatment of the subject was given by Lichnerowicz in his "Théorie Globale des connexions et des groupes d'holonomie"(1954 [27]). Thus the modern foundations of Finslerian geometry are best laid in the framework of fibre bundles as done in this book.

This book falls naturally into three parts:

I. Basics of Finselrian Geometry (Chapters I and II)
II. Classification of Finslerian manifolds (Chapters IV, V, VI)
III. Isometries, Projective and Conformal Transformations (Chapters III, VII, VIII)

CONTENTS

Chapter II
Finslerian Manifolds

Abstract

Chapter III
Isometries and affine vector fields on the unitary tangent fibre bundle

Abstract

Chapter VIII
Conformal vector fields on the unitary tangent fibre bundle

CHAPTER I

Linear Connections on a Space of Linear Elements

(**abstract**) Let M be a differentiable manifold of dimension n of class C∝. Let p: V(M) → M be the fibre bundle of non-zero tangent vectors to M , with fibre type Rn–{0} with structure group GL(n, R) the general linear group in n real variables. We denote by π: W(M) → M the fibre bundle of oriented directions tangent to M. Let E(M) be the linear fibre bundle of frames on M and $p^{-1}E(M)$ the induced fibre bundle of E(M) by p. An infinitesimal connection on $p^{-1}E(M)$ is called a linear connection of vectors ([1]). The study of this connection leads us to single out a condition of regularity (§5). In this case, independent tensor forms can be introduced on V(M). To a regular linear connection of vectors are associated canonically two torsion tensors S and T as well as three curvature tensors R, P and Q; we find expressions for them in (§7). In view of obtaining the formulas of the habitual linear connections we establish a reduction theorem (§8). With the help of covariant derivations of two types ∇ and $\dot{\nabla}$ we form three Ricci identities for a vector field in the large sense (§9). In §10 we show that there exist between the two torsion tensors S and T as well as among the three curvature tensors R, P and Q of a general regular connection five identities called Bianchi identities. We then give explicit formulas for them.

I. Regular Linear Connections

1. Fibre Bundles V(M) and W(M) over M.

a. Let M be a differentiable manifold of dimension n of class C^{∞}. The space V(M) of non-zero tangent vectors to M can be given the structure of differentiable fibre bundle over M of dimension 2n with structure group the linear group of n real variables GL(n, **R**), the fibre being isomorphic to the vector space $\mathbf{R}^n$ minus the origin. In the following a point of V(M) will be denoted by z = (x, v). We denote by p the canonical projection of each vector z of V(M) to its origin x∈ V(M) :

$$pz = x.$$

b. We call oriented tangent direction at a point x∈M the equivalence class defined on the non-zero vectors of the origin x by the positive collinearity: two non-zero vectors z_1 and z_2 ∈V(M) have the same direction if there existe a scalar λ>0 such that

$z_2=\lambda z_1$.The quotient space of V(M) by the equivalence relation defined by the positive collinearity will be called the *space of oriented tangent directions to* V(M) and will be denoted by W(M).The space V(M) is fibred over W(M) with the group of positive homothecies as the structure group.We denote a point of W(M) by y;and the canonical map of V(M) over W(M) by η.We have $\eta z=y$ .It is clear that W(M) is endowed with the structure of a differentiable fibre bundle over M, whose dimension is (2n-1) and whose fibres are homeomorphic to the ball Sn-1 and whose structure group is O(n). If π is the corresponding canonical projection, we have

$$\pi\, y = x$$

Between the maps p, η and π we have the relation

$$\pi \,.\, \eta = p$$

2. **Frames and Co-frames.**

a. Let E(M, F, G) be a differentiable fibre bundle over the base space M of fibre type F and of structure group G. If f denotes a differentiable mapping of a manifold M′ to the manifold M we know (Steenrod [23], p.47) that we can, starting with *E*(M, F, G), construct, a differentiable bundle E′(M′, F, G) with base M′ and of the same fibre types and the same structure group as E, this space is the induced fibre bundle from E by f and will be denoted by $f^{-1}E$. If one denotes respectively by g and g′ the canonical maps of E $\rightarrow$ M and of E′ $\rightarrow$ M′ we have the following commutative diagram:

$$\begin{array}{ccc} E' = f^{-1}E & \xrightarrow{h} & E \\ g'\downarrow & & \downarrow g \\ M' & \xrightarrow{f} & M \end{array}$$

where h is the *induced map* from *E'* $\rightarrow$ *E* and f.g′= g.h

b. Let E(M) be the space of principal fibre bundles of linear frames over M of fibre type and structure group the linear group Gl(n,**R**). The fibre bundle, induced from E(M) by π: W(M) $\rightarrow$ M, (resp. by p:V(M)$\rightarrow$M) is then a principal fibre bundle $\pi^{-1}E(M)$ (resp. $p^{-1}E(M)$) with base W(M) (resp. V(M)) of fibre type and structure group identical to GL(n, **R**) and will be called *the fibre bundle over W (resp. over V) of linear frames*. It is clear that $p^{-1}E(M)$ is none other than the fibre bundle induced from $\pi^{-1}E(M)$ by: η:V(M) $\rightarrow$ W(M). To construct these spaces one can also proceed with the usual method of coverings.

c. Let us consider a covering of M by open and connected neighbourhoods (U,V,...). At each point $z \in p^{-1}(U)$ we consider a frame $R^z{}_U$, depending differentiably on z: it is a basis of T_{pz}, that is to say an ordered set of n linearly independent vectors of T_{pz},. If U and V are two neighbourhoods of the covering, let $R^z{}_U$ and $R^z{}_V$ be the frames attached toz $\in p^{-1}(U\cap V)$, then there exists a regular matrix element of the group GL(n, **R**) such that

$$R^z{}_V = R^z{}_U\, A^U{}_V(z), \quad z \in p^{-1}(U\cap V),$$

To such a frame of T_{pz} corresponds in $T^*{}_{pz}$ a dual basis or co-frame $\alpha^U z$: it is an ordered set of n linear forms $(\alpha^1,\alpha^2, \ldots \alpha^n)$ linearly independent in $T^*{}_{pz}$. If $\alpha^U z$ and $\alpha^V z$ are dual co-frames of $R^z{}_U$ and $R^z{}_V$ we have, in matrix notation, the relation

$$\alpha^U z = A^U{}_V\, \alpha^V z \ \text{ where } z \in p^{-1}(U\cap V),$$

Let Q^z be the set of frames attached to $z \in p^{-1}(U\cap V)$, over the union

$$\bigcup_{z\in V} Q^z \tag{2.1}$$

one can define a topology and a natural structure of a differentiable manifold: to every system of local coordinates (x^i) of M we let correspond a system of local coordinates for (2.1), where the coordinates of an element of $R^z \in \bigcup_{z\varepsilon V} Q^z$ are defined by the coordinates (x^i, v^i) of its origin z, the (v^i) being the components of the vector z with respect to the natural frame of the origin x, and by the matrix A defining R^z (with respect to the same natural frame). The projection which makes correspond to every element $R^z \in \cup_{z\in V} Q^z$ its origin z defines $\cup_{z\in V} Q^z$ as a principal fibre bundle over V(M) with GL(n,**R**) as the structure group. This principal fibre bundle will be identified with $p^{-1}E(M)$.

3.Tensors and Tensor Forms.

a. Tensor fields in the large sense. We call an *affine tensor field in the large sense a* map which lets correspond to every $z \in V(M)$ an element of the affine tensor algebra constructed over T_{pz}. By abuse of language we call a tensor in the large sense instead of a tensor field in the large sense.

b. Tensor fields in the restricted sense. Let t be tensor field in the large sense. To two points z_1 and z_2 belonging to $\eta^{-1}(y)$ correspond two values $t(z_1)$ and $t(z_2)$ in general not proportional. Let us suppose that for $z_2 = \lambda z_1$. We have necessarily $t(z_2) = f(\lambda)\, t(z_1)$, $(\lambda > 0)$, where $f(\lambda)$ is a continuous function of λ. Let us write this relation for three points of $\eta^{-1}(y)$ say z_1, $z_2 = \lambda z_1$, $z_3 = \mu z_2$ we obtain

$$t(z_3) = f(\mu)\, t(z_2) = f(\mu) f(\lambda)\, t(z_1),$$

Since we have $z_3 = \mu\lambda z_1$ we also have

$$t(z_3) = f(\mu\lambda)\, t(z_1)$$

It follows that f must satisfy

$$f(\mu\lambda) = f(\mu) f(\lambda)$$

The only continuous solution of this functional equation is of the form
([9], p.19):

$$f(\lambda) = \lambda^p$$

where $p \in \mathbf{R}$. We call a *tensor field in the restricted sense* (by abuse of language a restricted tensor) a tensor field t such that if z_1 and $z_2 \in \eta^{-1}(y)$, we have for $z_2=\lambda z_1$

$$t(z_2) = \lambda^p \, t(z_1) \tag{3.2}$$

p will be called the *degree* of the tensor t. In particular, a tensor field of degree zero in the restricted sense is nothing else but a tensor field over W(M).

C. By a tensorial r-form in the large sense over V(M) of definite type (k, l) we understand a map which lets corrrespond to every element $z \in V(M)$ an element of $T^k_{pz} \otimes T^{*l}_{pz} \otimes \Lambda^{(r)}_{pz}$ where $\Lambda^{(r)}_{pz}$ is the space at z of r-forms with scalar values of V(M). Such a form is a tensorial r-form of the usual type of the fibre bundle $p^{-1}E(M)$, λ being a given real constant > 0. Let μ_λ be the transformation of V defined by $\mu_\lambda : z \to \lambda z$ a tensorial r-form τ in the restricted sense of homogeneous degree p is a tensorial r-form over V(M) such that for every λ we have

$$\mu^*_\lambda \tau = \lambda^p \tau$$

4. Linear Connections. Let $p^{-1}E(V(M))$ be the principal fibre bundle of linear frames on V(M). An infinitesimal connection of $p^{-1}E(M)$ will be called a *linear connection of vectors* [1], [1c]. Given a covering of V(M) by neighbourhoods with local sections of $p^{-1}E(M)$ a linear connection of vectors can be given by the data in each U of a 1-differential form ω^z_U over $p^{-1}(U)$ with values in the Lie algebra of the group GL(n, **R**) and such that for $z \in p^{-1}(U \cap V)$, we have

$$R^z_V = R^z_U \, A^U_V(z), \qquad A \in GL(n, \mathbf{R})$$

they satisfy the conditions of coherence

(4.1) $\omega^z{}_V = A^{-1U}{}_V\, \omega^z{}_U\, A^U{}_V + A^{-1U}{}_V\, d\, A^U{}_V$

where in each U endowed with frames the form $\omega^z{}_U$ is represented by an (nxn) matrix whose elements are the 1-differential forms on $p^{-1}(U)$: the elements of the matrices $\omega^z{}_U$ and $\omega^z{}_V$ will be denoted by

$$\omega^z{}_U = (\omega^i{}_j), \quad \omega^z{}_V = (\omega^{k'}{}_{h'})$$

To the linear connection thus envisaged corresponds a one form over V(M) with values in the Lie algebra GL(n,**R**). *An infinitesimal connection of* $\pi^{-1}E(M)$ *will be called a linear connection of directions.* Such a connection can be defined in a manner similar to the preceding.

5. Absolute differential in a linear connection. Regular linear connection.

a. Given a covering of M by neighbourhoods with frames, let us consider a vector field W in the large sense in V (M). For $z \in p^{-1}(U)$, it can be defined by one row matrix of its components W^U. The absolute differential of W relative to the linear connection of vectors is

(5.1) $\nabla W^U = dW^U + \omega_U W^U$

where ∇W^U defines a linear differential form over V(M) with vector values. Similarly, the absolute differential of a covariant vector β_U is defined by

(5.2) $\nabla\beta_U = d\beta_U - \beta_U\, \omega_U$

There exists a canonical vector field in the large sense: one which to every z of V(M) lets correspond the vector T_{pz} *defined by z. This canonical vector field will be denoted in the following by v. From the consideration of its absolute differential we deduce the local n-forms*

(5.3) $$\theta^i = \nabla v^i = dv^i + \omega^i_j v^j$$

which define a one form θ of vector type. The set (α^i, dv^i) form a co-frame of the vector space Θ_z tangent to V(M) at z ; in order that the set (α^i, θ^i) forms a co-frame of Θz it is necessary and sufficient that the system of n-forms to which are reduced the θ^i for $\alpha^i=0$, (i=1, 2, .., n) be linearly independent. Let us suppose that the linear connection of vectors is defined with respect to a local section of E(M) by:

(5.4) $$\omega^i_j = \Gamma^i_{jk}\, dx^k + C^i_{jk}\, dv^k.$$

Using the coherence condition we see that C^i_{jk} are the components of a tensor of type (1, 2); if we denote by μ^i the restrictions of θ^i to the fibre $p^{-1}(x)$ of (5.3) and (5.4) it follows that

(5.5) $$\mu^i = dv^i + v^j C^i_{jk}\, dv^k = (\delta^i_k + v^j C^i_{jk})\, dv^k$$

We are thus led to study the system in dv defined by $\mu^i=0$; in order that this system admits only the zero solution it is necessary and sufficient that the matrix

(5.6) $(\delta^i_k + v^j C^i_{jk})$, δ^i_k being the Kronecker symbol

be *regular* ; in this case the set (α^i, θ^i) forms effectively a co-frame of Θ_z. It is clear that this matrix defines a tensor of type (1,1).

Definition 1.5.- *A linear connection of vectors is called regular if the set* $(\alpha^i, \theta^i = \nabla v^i)$ *forms a co-frame of the vector space tangent to V(M) at* $z \in V(M)$.

b. In the following we suppose that the linear connection of vectors is regular; with respect to the co-frame (α^i, θ^i) the matrix ω^i_j is written

(5.7) $$\omega^i_j = \gamma^i_{jk}\, \alpha^k + C^i_{jk}\, \theta^k$$

ω being regular we deduce by putting (5.7) in (5.3) that the matrix

(5.8) $$L^i_k = \delta^i_k - v^j C^i_{jk}$$

is regular. In the following we will denote by M the inverse of L

$$L.M = M.L = I$$

where I is the identity matrix; thanks to the condition of coherence (4.1) the coefficients of two kinds of linear connection of vectors get transformed according to the formulas

(5.9) $$\gamma^{k'}{}_{h'l'} = A^{k'}{}_i A^j{}_{h'} A^k{}_{l'} \gamma^i{}_{jk} + A^{k'}{}_r \partial_{l'} A^r{}_{h'},$$

(5.10) $$C^{k'}{}_{h'l'} = A^{k'}{}_i A^j{}_{h'} A^k{}_{l'} C^i{}_{jk} + A^{k'}{}_r \dot{\partial}_{l''} A^r{}_{h'}$$

where $\partial_{l'}$ and $\dot{\partial}_{l''}$ denote respectively the Pfaffian derivatives with respect to $\alpha^{l'}$ and $\theta^{l'}$.

If we consider a covering of M endowed with the local sections of E(M) the $A^r{}_h$ depend only on x therefore $\dot{\partial}_{l'} A^r{}_{h'} = 0$, the relation (5.10) becomes

(5.11) $$C^{k'}{}_{h'l'} = A^{k'}{}_i A^j{}_{h'} A^k{}_{l'} C^i{}_{jk}$$

Thus it follows that relative to the co-frames of such a covering, the $C^i{}_{jk}$ are the components of a tensor T of type (1,2). Thus with respect to the coframe (dx^i, θ^i) the matirx $\omega^i{}_j$ of the linear connection of vectors can be written

(5.12) $$\omega^i{}_j = \overset{*}{\Gamma}{}^i{}_{jk}\, dx^k + T^i{}_{jk}\theta^k$$

(5.12) is identical to (5.4); in virtue of (5.3), we have

$$\overset{*}{\Gamma}{}^{i}_{jk}dx^{k}+T^{i}_{jk}\theta^{k}=\Gamma^{i}_{jk}dx^{k}+C^{i}_{jh}[\theta^{h}-v^{r}(\overset{*}{\Gamma}{}^{h}_{rk}dx^{k}+T^{h}_{rk}\theta^{r})]$$

hence

(5.13) $$\overset{*}{\Gamma}{}^{i}_{jk} = \Gamma^{i}_{jk} - C^{i}_{jh}\, v^{r}\, \overset{*}{\Gamma}{}^{h}_{rk}$$

(5.14) $$T^{i}_{jk}= C^{i}_{jk}- C^{i}_{jh}\, v^{r}\, T^{h}_{rk}$$

It is clear that for the co-frames (dx^i, θ^i), the $L^i_k = \delta^i_k - v^j T^i_{jk}$ are the components of a tensor of type (1,1); it is the same for the matrix M, the inverse of L defined by (5.6).

6. Exterior Differential Forms.

a. Given a regular linear connection of vectors, we have shown in the previous paragraph that the set (α^i, θ^i) constitutes a basis for the 1-forms over V(M). Thus any 1-form over V(M) can be written

(6.1) $$\pi = a_i\, \alpha^i + b_i \theta^i$$

For π to be the inverse image over V(M) of a 1-form overW(M) by the canonical map $\eta : V(M) \to W(M)$ it is necessary and sufficient that whatever be z_1 and $z_2 \in \eta^{-1}(y)$ one must have $z_2 = \lambda z_1$, $(\lambda > 0)$

(6.2) $$\pi(z_2) = \pi(z_1)$$

From $v_2 = \lambda v_1$ we obtain

$$\nabla v^i_2 = \lambda \nabla v^i_1 + v^i_1 d\lambda$$

Putting this relation in (6.2) and on identifying the two parts we have

(6.3) $$a_i(z_2) = a_i(z_1)$$
$$b_i(z_2) = \lambda^{-1}\, b_i\,(z_1)$$
$$b_i\, v^i = 0$$

In the following, the operation of multiplication contracted by v will be denoted by the index 0; thus the last relation of (6.3) can be written

$$b_0 = 0$$

Similarly given a differential 2-form on V(M)

(6.4) $$\frac{1}{2} a_{ij}\alpha^i \wedge \alpha^j + b_{ij}\, \alpha^i \wedge \theta^j + \frac{1}{2} C_{ij}\, \theta^i \wedge \theta^j$$

In order that it be the inverse image on V(M) of a differential 2-form on W(M) by η (V → W) it is necessary and sufficient that we should have for $z_2 = \lambda z_1$ ($\lambda>0$)

(6.5) $$\begin{aligned} &a_{ij}(z_2) = a_{ij}(z_1) \\ &b_{ij}(z_2) = \lambda\text{-}1\, b_{ij}(z_1) \qquad b_{io} = 0 \\ &C_{ij}(z_2) = \lambda\text{-}2C_{ij}(z_1),\ C_{oj} = 0 = C_{jo} \end{aligned}$$

In a general manner, *for an r-form on V(M) to be the inverse image of an r-form on W(M) by the canonical map η: V(M)→ W(M) it is necessary and sufficient that the coefficient of the mixed term containing* α^i, *p times and* θ^i, *q times, p+ q = r be restricted of degree -q satisfying the condition of homogeneity of multplication contracted by v.* In this case this r-form will be identified with an r-form on W(M). The preceding result applies to the connection form:

For a regular linear connection ω on V(M) to be the inverse image of a linear connection on W(M) by η:V(M) → W(M) it is necessary and sufficient that for z_1 and $z_2 = \lambda z_1$, ($\lambda > 0$), $\in \eta\text{-}1\,(y)$ we have

(6.6) $$\begin{aligned} &\gamma^i_{jk}(z_2) = \gamma^i_{jk}(z_1) \\ &C^i_{jk}(z_2) = \lambda^{-1} C^i_{jk}(z_1) \\ &C^i_{jo} = 0 \end{aligned}$$

and we will identify ω with a linear connection of directions. It is clear that the condition of homogeneity of the coefficients of the second kind of the connection is invariant by a change of frames.

Let t be a tensor in the large sense on V(M) of type (1, 1), the absolute differential of t relative to the linear connection of vectors is

$$\nabla t^i_j = dt^i_j + \omega^i_r t^r_j - t^i_r \omega^r_j.$$

In what follows we denote by ∇_k and $\nabla_{\dot{k}}$ respectively the covariant derivatives with respect to the co-frames α^k and θ^k; thus in virtue of (5.7), the above relation gives us

$$\nabla_k t^i_j = \partial_k t^i_j + \gamma^i_{rk} t^r_j - t^i_r \gamma^r_{jk}$$

$$\nabla^{\bullet}{}_k t^i_j = \dot{\partial}_k t^i_j + C^i_{rk} t^r_j - t^i_r C^r_{jk}.$$

In particular we have

$$\nabla_k v^i = 0 \; ; \; \nabla_{\dot{k}} v^i = \delta^i_k$$

In earliers sections we have introduced the Pfaffian derivatives with respect to the co-frames (α^k, θ^k). In the following, we are thus led to reason in terms of the co-frame (α^k, dv^k). Denoting by δ_k and $\dot{\delta}_k$ the Pfaffian derivatives relative to (α^k, dv^k) we propose to establish the relations that exist among these different derivations. To do this we substitute the expression for ω^i_j defined by (5.7) in (5.3),

$$\theta^k = dv^k + (\gamma^k_{ih}\,\alpha^h + C^k_{ih}\,\theta^h)\,v^i \tag{6.7}$$

Let Φ be a function with real values in the large sense on V(M) at a point $z \in V(M)$, its differential can be written indifferently

$$d\Phi = \partial_k \Phi \alpha^k + \dot{\partial}_k \Phi \theta^k = \delta_k \Phi \alpha^k + \delta_{\dot{k}} \Phi dv^k$$

In virtue of (6.7), the above relation becomes

$$\partial_k \Phi\, \alpha^k + \partial^{\cdot}_k \Phi\, \theta^k = (\delta_k \Phi - \delta^{\cdot}_r \Phi\, \gamma^r_{ok})\alpha^k + (\delta^{\cdot}_k \Phi - \delta^{\cdot}_r \Phi\, C^r_{ok})\, \theta^k$$

Thus

(6.8) $$\partial_k = \delta_k - \gamma^r_{ok}\, \delta^{\cdot}_r ,$$

(6.9) $$\partial^{\cdot}_k = \delta^{\cdot}_k - C^r_{ok}\, \delta^{\cdot}_r$$

Thus the derivation operators ∂ and $\partial^{\cdot}$ are expressed by means of δ and $\delta^{\cdot}$ and the coefficients of the connection by (6.8) and (6.9), the linear connection in question being regular the preceding formulas are invertible and we have

(6.10) $$\delta_k = \partial_k + \gamma^h_{ok} M^r_h\, \partial^{\cdot}_r ,$$

(6.11) $$\delta^{\cdot}_k = M^r_k\, \partial^{\cdot}_r$$

b. Let us denote by TV(M) the fibre bundle of tangent vectors to V(M), and by ρ the canonical linear map $TV(M) \rightarrow p^{-1}T(M)$. This means that if $\hat{X} \in TV(M)$, we have $\rho(\hat{X}) = (z, p_*(\hat{X}))$, $z \in V(M)$ et p_* is the linear tangent map. Let V_z be the set of vertical vectors at $z \in V(M)$, that is to say, vectors that are tangent to the fibre passing through $z \in V(M)$ ($V_z = \ker p_*$ where $p_*: TV(M) \rightarrow T(M)$). Let ∇ be a covariant derivation in the fibre bundle $p^{-1}TM \rightarrow V(M)$. We now define a linear map

$$\mu_z : T_z V(M) \rightarrow T_{pz}(M)$$
$$\mu_z(\hat{X}) = \nabla_{\hat{X}}\, v$$

where $\hat{X} \in TV(M)$

Now we show that if the connection is regular the linear map μ_z defines an isomorphism of V_z on T_{pz}.for each $z \in V(M)$.

Taking a vertical basis $\delta^{\cdot}_j = \dfrac{\delta}{\delta v^j}$ of V_z, we have

$$\mu_z\left(\frac{\delta}{\delta v^j}\right) = \nabla_{\delta_j}(v^i\,\delta_i) = (\delta^i_j + v^r\,C^i_{rj})\,\delta_i$$

Now the matrix in the bracket of the right hand side is regular. So μ_z defines an isomorphism of V_z onto T_{pz}. Let us suppose that the connection is regular and let $H_z \subset T_zV(M)$ be the kernel of μ_z. Then at each point $z \in V(M)$ we have the following decomposition

$$T_zV(M) = H_z \oplus V_z, \qquad H_z = \ker \mu_z, \qquad H_z \cap V_z = 0.$$

where H_z (respectively V_z) is the horizontal (respectively vertical) space defined by ∇. Every tangent vector $\hat{X} \in TV(M)$ decomposes into

$$\hat{X} = H\hat{X} + V\hat{X}, \qquad H\hat{X} \in H_z,\ V\hat{X} \in V_z$$

II. ---Curvature and torsion of a regular linear connection

7. Torsion and curvature tensors of a general linear connection.

i. Torsion tensors. Let ω be a regular linear connection of vectors represented in each U by the matrix ω^i_j defined by (5.7) and satisfying the condition of coherence. It is clear that the one row matrices whose elements are

$$\Sigma^i = d\alpha^i + \omega^i_j \wedge \alpha^j \tag{7.1}$$

define on V(M) a 2-form with vector values. We call this vector 2-form the torsion form associated to the linear connection. Since the Pfaffian system $\alpha^i = 0$, is completely integrable, the elements $d\alpha^i$ can be written

$$d\alpha^i = \frac{1}{2}\, b^i_{jk}\, \alpha^j \wedge \alpha^k + a^i_{jk}\, \alpha^j \wedge \theta^k \tag{7.2}$$

where the b^i_{jk} are skew-symmetric with respect to the lower indices; taking into account this relation, the torsion form can be put in the form

(7.3) $\Sigma^i = d\alpha^i + \omega^i_j \wedge \alpha^j = \frac{1}{2} S^i_{jk}\, \alpha^j \wedge \alpha^k + T^i_{jk}\, \theta^k \wedge \alpha^j$

from which we have

$$S^i_{jk} + S^i_{kj} = 0$$

where the S^i_{jk} and T^i_{jk} are the components of two tensors of type (1, 2) which will be called *torsion tensors* of the two kinds of the connection. On substituting (7.2) in (7.3) and on identifying the two sides, we obtain

(7.4) $$S^i_{jk} = b^i_{jk} - (\gamma^i_{jk} - \gamma^i_{kj})$$

(7.5) $$T^i_{jk} = C^i_{jk} - a^i_{jk}$$

ii. Curvature tensors.

a. To the regular linear connection ω on V(M) we can associate a 2-tensorial form Ω of type adj g^{-1} over V(M) having for components

(7.6) $$\Omega^i_j = d\omega^i_j + \omega^i_r \wedge \omega^r_j$$

We call this 2-form ***the curvature form*** of the connection. With respect to the co-frames, the Ω^i_j can be written

(7.7) $$\Omega^i_j = \frac{1}{2} R^i_{jkl}\alpha^k \wedge \alpha^l + P^i_{jkl}\alpha^k \wedge \theta^l + \frac{1}{2} Q^i_{jkl}\theta^k \wedge \theta^l;$$

the tensors R, P and Q will be called *the curvature tensors* of the connection, the first and the last being skew-symmetric with respect to the indices k and l. In order to evaluate these tensors it is necessary and sufficient to develop the right hand side of (7.6) and to identify it to (7.7). To do this we obtain by differentiation

$$(7.8)\quad d\omega^i_j=\frac{1}{2}(\partial_k\gamma^i_{jl}-\partial_l\gamma^i_{jk})\alpha^k\wedge\alpha^l+(\partial_k C^i_{jl}-\dot\partial_l\gamma^i_{jk})\alpha^k\wedge\theta^l$$
$$+\frac{1}{2}(\dot\partial_k C^i_{jl}-\dot\partial_l C^i_{jk})\theta^k\wedge\theta^l+\gamma^i_{jr}d\alpha^r+C^i_{jr}d\theta^r$$

In the above relation we see appear the differentials of the co-frames; $d\alpha^r$ and $d\theta^r$ the first being calculated from (7.3); as to the second we have by (5.3)

$$(7.9)\qquad d\theta^r = d\omega^r_s\, v^s - \omega^r_s\wedge dv^s = \theta^s\wedge\omega^r_s+\Omega^r_o$$

On the other hand, the last term of the right hand side of (7.6) becomes

$$(7.10)\quad \omega^i_r\wedge\omega^r_j=\frac{1}{2}(\gamma^i_{rk}\gamma^r_{jl}-\gamma^i_{rl}\gamma^r_{jk})\alpha^k\wedge\alpha^l+(\gamma^i_{rk}C^r_{jl}-\gamma^r_{jk}C^i_{rl})\alpha^k\wedge\theta^l$$
$$+\frac{1}{2}(C^i_{rk}C^r_{jl}-C^i_{rl}C^r_{jk})\theta^k\wedge\theta^l$$

Adding (7.8) to (7.10) and identifying the expression thus obtained to (7.7) we have

$$(7.11)\qquad R^i_{jkl}=\overset{*}{R}{}^i_{jkl}+C^i_{jr}R^r_{okl}$$

$$(7.12)\qquad P^i_{jkl}=\overset{*}{P}{}^i_{jkl}+C^i_{jr}P^r_{okl}$$

$$(7.13)\qquad Q^i_{jkl}=\overset{*}{Q}{}^i_{jkl}+C^i_{jr}Q^r_{okl}$$

where we have put

$$(7.14)\qquad \overset{*}{R}{}^i_{jkl}=(\partial_k\gamma^i_{jl}-\partial_l\gamma^i_{jk})+(\gamma^i_{rk}\gamma^r_{jl}-\gamma^i_{rl}\gamma^r_{jk})+\gamma^i_{jr}(S^r_{kl}+\gamma^r_{kl}-\gamma^r_{lk})$$

$$(7.15)\qquad \overset{*}{P}{}^i_{jkl}=(\partial_k C^i_{jl}-\dot\partial_l\gamma^i_{jk})$$
$$+(\gamma^i_{rk}C^r_{jl}-\gamma^r_{jk}C^i_{rl}-\gamma^r_{lk}C^i_{jr})+\gamma^i_{jr}(C^r_{kl}-T^r_{kl})$$

$$(7.16)\qquad \overset{*}{Q}{}^i_{jkl}=(\dot\partial_k C^i_{jl}-\dot\partial_l C^i_{jk})+(C^i_{rk}C^r_{jl}-C^i_{rl}C^r_{jk})$$
$$+C^i_{jr}(C^r_{kl}-C^r_{lk})$$

Multiplying the two sides of (7.11) by v^j we get

$$(\delta^i_s - C^i_{os})\, R^s_{okl} = \overset{*}{R}{}^i_{okl}$$

The expression within the brackets is none other than the matrix L^i_s defined by (5.8); the linear connection involved being regular, it follows, on multiplying the two sides by M^i_r that

(7.17) $$R^i_{okl} = M^i_r \overset{*}{R}{}^r_{okl}$$

Thus (7.11) become

(7.18) $$R^i_{jkl} = \overset{*}{R}{}^i_{jkl} + C^i_{jr}\, M^r_a\, \overset{*}{R}{}^a_{okl}$$

In the same way, for the other two tensors we obtain

(7.19) $$P^i_{jkl} = \overset{*}{P}{}^i_{jkl} + C^i_{jr}\, M^r_a\, \overset{*}{P}{}^a_{okl}$$

(7.20) $$Q^i_{jkl} = \overset{*}{Q}{}^i_{jkl} + C^i_{jr}\, M^r_a\, \overset{*}{Q}{}^a_{okl}$$

where R*, P*, Q* are defined by (7.14), (7.15) and (7.16). Thus we see that *three curvature tensors of a regular linear connection are expressed in terms of the coefficients of the connection and their first derivatives as well as torsion tensors by the formulas (7.18), (7.19) and* (7.20).

b. It is often convenient to reason in terms of the co-frames (dx^i, θ^i) of local coordinates. In this case the matrix of the linear connection is represented by (5.12), by (7.2) it follows that

(7.21) $$b^i_{jk} = 0, \quad a^i_{jk} = 0$$

From (7.4) we obtain

(7.22) $$S^i_{jk} = -(\dot{\Gamma}^i_{jk} - \dot{\Gamma}^i_{kj})$$

and the coefficients of the second kind C^i_{jk} coincide with the torsion tensor T^i_{jk}. The formulas (7.14), (7.15) and (7.16) become in this case

$$\overset{*}{R}{}^i_{jkl} = \partial_k \dot{\Gamma}^i_{jl} - \partial_l \dot{\Gamma}^i_{jk} + \dot{\Gamma}^i_{rk} \dot{\Gamma}^r_{jl} - \dot{\Gamma}^i_{rl} \dot{\Gamma}^r_{jk} \tag{7.23}$$

$$\overset{*}{P}{}^i_{jkl} = \nabla_k T^i_{jl} - \partial^\bullet_l \dot{\Gamma}^i_{jk} \tag{7.24}$$

$$\overset{*}{Q}{}^i_{jkl} = \nabla^\cdot_k T^i_{jl} - \nabla^\cdot_l T^i_{jk} + T^i_{rl} T^r_{jk} - T^i_{rk} T^r_{jl} \tag{7.25}$$

and the curvature tensors become

$$R^i_{jkl} = \overset{*}{R}{}^i_{jkl} + T^i_{jr} M^r_a \overset{*}{R}{}^a_{okl} \tag{7.26}$$

$$P^i_{jkl} = \overset{*}{P}{}^i_{jkl} + T^i_{jr} M^r_a \overset{*}{P}{}^a_{okl} \tag{7.27}$$

$$Q^i_{jkl} = \overset{*}{Q}{}^i_{jkl} + T^i_{jr} M^r_a \overset{*}{Q}{}^a_{okl} \tag{7.28}$$

By the preceding formulas it is clear that $\overset{*}{R}$, $\overset{*}{P}$, and $\overset{*}{Q}$ define in this case three tensors.

8. Particular case of a linear connection of directions. Conditions of reduction.

a. In the case of regular linear connection of directions we know that the coefficients of ω^i_j satisfy (6.6), the preceding formulas giving the torsion and curvature tensors of a general linear connection are in particular valid for such a connection, and we have in addition

$$T^i_{jo} = 0, \;\; P^i_{jko} = 0, \;\; Q^i_{jol} = Q^i_{jko} = 0 \tag{8.1}$$

b. Let ω be a regular linear connection on $p^{-1}E(M)$; let us suppose that it is the inverse image of linear connection on $E(M)$ by the canonical map of $p^{-1}E(M)$ on $E(M)$ with respect to the co-frame (dx^i, dv^i) the matrix of this can be written

(8.2) $$\omega^i_j = \Gamma^i_{jk}(x)dx^k ,$$

It follows , by (5.13) and (5.14)

$$\overset{*}{\Gamma}{}^i_{jk} = \Gamma^i_{jk}(x), \qquad T^i_{jk} = 0$$

In virtue of (7.27), (7.24) and (6.9), we have

$$P^i_{jkl} = \overset{*}{P}{}^i_{jkl} = -\partial^{\bullet}_l \overset{*}{\Gamma}{}^i_{jk} = -\delta^{\bullet}_l \Gamma^i_{jk} = 0$$

Thus the tensors T and P are zero. Conversely, let us suppose that the tensors T and P of the regular linear connection on $p^{-1}E(M)$ are zero; from the relations (5.13), (5.14), (6.9), (7.27) and (7.24) it follows that C^i_{jk} is identically zero and the $\overset{*}{\Gamma}{}^i_{jk} = \Gamma^i_{jk}$ do not depend on the direction, and the matrix of the linear connection is of the form (8.2). Thus we have:

Theorem. *In order that a regular linear connection ω on $p^{-1}E(M)$ be the reciprocal image of a linear connection on E(M) by the canonical map of $p^{-1}E(M)$ over E(M) it is necessary and sufficient that the tensors T= 0 and P = 0.*

If it is so, Q = 0 and the connection can be identified with a linear connection on E(M).

9. Ricci Identities. –Take a covering of M by neighbourhoods (U) endowed with a natural frame of local coordinates. Let X be a vector field in the large sense; in each U, X defines a 0-form with vector values $X_z^U = (X^i)$; the relation $ddX^i = 0$ becomes

$$\frac{1}{2}(\partial_k\partial_l - \partial_l\partial_k)X^i dx^k \wedge dx^l + (\partial_k\dot{\partial}_l - \dot{\partial}_l\partial_k)X^i dx^k \wedge \theta^l$$

$$+ \frac{1}{2}(\dot{\partial}_k\dot{\partial}_l - \dot{\partial}_l\dot{\partial}_k)X^i\theta^k \wedge \theta^l + \dot{\partial}_r X^i d\theta^r = 0$$

where ∂_k and $\dot{\partial}_k$ the Pfaffian derivatives with respect to the local co-frame (dx^k, θ^k). Taking into account (7.9) the above relation gives us

$$(9.1) \qquad (\partial_k \partial_l - \partial_l \partial_k) X^i + \dot{\partial}_r X^i R^r{}_{okl} = 0$$

$$(9.2) \qquad (\partial_k \dot{\partial}_l - \dot{\partial}_l \partial_k) X^i + \dot{\partial}_r X^i (P^r{}_{okl} - \overset{*}{\Gamma}{}^r{}_{lk}) = 0$$

$$(9.3) \qquad (\dot{\partial}_k \dot{\partial}_l - \dot{\partial}_l \dot{\partial}_k) X^i + \dot{\partial}_r X^i (Q^r{}_{okl} + T^r{}_{kl} - T^r{}_{lk}) = 0$$

On the other hand the covariant derivation of type ∇_l of X is

$$\nabla_l X^i = \partial_l X^i + X^s \overset{*}{\Gamma}{}^i{}_{sl}$$

A second derivation of the same type gives us

$$\begin{aligned} \nabla_k \nabla_l X^i = {} & \partial_k \partial_l X^i + \partial_k X^s \overset{*}{\Gamma}{}^i{}_{sl} + X^s \partial_k \overset{*}{\Gamma}{}^i{}_{sl} \\ & + \partial_l X^r \overset{*}{\Gamma}{}^i{}_{rk} + X^s \overset{*}{\Gamma}{}^r{}_{sl} \overset{*}{\Gamma}{}^i{}_{rk} - \nabla_r X^i \overset{*}{\Gamma}{}^r{}_{lk} \end{aligned}$$

From this we deduce the identity, on taking into account (9.1), (7.22) and (7.23),

$$(9.4) \qquad (\nabla_k \nabla_l - \nabla_l \nabla_k) X^i = X^r R^i{}_{rkl} - \dot{\nabla}_r X^i R^r{}_{okl} - \nabla_r X^i S^r{}_{kl}$$

In an analogous manner we obtain

$$(9.5) \qquad (\nabla_k \dot{\nabla}_l - \dot{\nabla}_l \nabla_k) X^i = X^r P^i{}_{rkl} - \dot{\nabla}_r X^i P^r{}_{okl} + \nabla_r X^i T^r{}_{kl}$$

$$(9.6) \qquad (\dot{\nabla}_k \dot{\nabla}_l - \dot{\nabla}_l \dot{\nabla}_k) X^i = X^r Q^i{}_{rkl} - \dot{\nabla}_r X^i Q^r{}_{okl}$$

More generally, let $t^{i_1 \ldots i_\alpha}{}_{j_1 \ldots j_\beta}$ be a tensor field in the large sense; we obtain

$$\begin{aligned} (9.7) \quad (\nabla_k \nabla_l - \nabla_l \nabla_k) t^{i_1 \ldots i_\alpha}{}_{j_1 \ldots j_\beta} = {} & \sum_{\gamma=1}^{\alpha} R^{i_\gamma}{}_{rkl} t^{i_1 \ldots r \ldots i_\alpha}{}_{j_1 \ldots j_\beta} \\ & - \sum_{\mu=1}^{\beta} R^r{}_{j_\mu kl} t^{i_1 \ldots\ldots i_\alpha}{}_{j_1 \ldots r \ldots j_\beta} - R^r{}_{okl} \dot{\nabla}_r t^{i_1 \ldots i_\alpha}{}_{j_1 \ldots j_\beta} - S^r{}_{kl} \nabla_r t^{i_1 \ldots i_\alpha}{}_{j_1 \ldots j_\beta} \end{aligned}$$

(9.8) $$(\nabla_k \dot\nabla_l - \dot\nabla_l \nabla_k) t^{i_1 \ldots i_\alpha}{}_{j_1 \ldots j_\beta} = \sum_{\gamma=1}^{\alpha} P^{i\gamma}{}_{rkl}\, t^{i_1 \ldots r \ldots i_\alpha}{}_{j_1 \ldots j_\beta}$$

$$- \sum_{\mu=1}^{\beta} P^r{}_{j\mu kl}\, t^{i_1 \ldots\ldots i_\alpha}{}_{j_1 \ldots r \ldots j_\beta} - P^r{}_{okl} \dot\nabla_r t^{i_1 \ldots i_\alpha}{}_{j_1 \ldots j_\beta} + T^r{}_{kl} \nabla_r t^{i_1 \ldots i_\alpha}{}_{j_1 \ldots j_\beta}$$

(9.9) $$(\dot\nabla_k \dot\nabla_l - \dot\nabla_l \dot\nabla_k) t^{i_1 \ldots i_\alpha}{}_{j_1 \ldots j_\beta} = \sum_{\gamma=1}^{\alpha} Q^{i\gamma}{}_{rkl}\, t^{i_1 \ldots r \ldots i_\alpha}{}_{j_1 \ldots j_\beta}$$

$$- \sum_{\mu=1}^{\beta} Q^r{}_{j\mu kl}\, t^{i_1 \ldots\ldots i_\alpha}{}_{j_1 \ldots r \ldots j_\beta} - Q^r{}_{okl} \dot\nabla_r t^{i_1 \ldots i_\alpha}{}_{j_1 \ldots j_\beta}$$

To the identities of (9.7), (9.8) and (9.9) we give the name the **Ricci identities [1]**.

10. Bianchi Identities [1]. We have associated to the regular linear connection on V(M) three curvature tensors and two torsion tensors; among these different tensors and their covariant derivatives there exist relations which we now are going to establish. By exterior differentiation from the formulas (7.1), and (7.6) we obtain

(10.1) $$d\Sigma^i = \Omega^i_j \wedge \alpha^j - \omega^i_j \wedge \Sigma^j$$

(10.2) $$d\Omega^i_j = \Omega^i_s \wedge \omega^s_j - \omega^i_s \wedge \Omega^s_j$$

The relations (10.1) and (10.2) are called the **Bianchi identities**. On identifying in (10.1) the terms in $\alpha^k \wedge \alpha^l \wedge \alpha^m$ we obtain

(10.3) $$\mathbf{S}\, R^i{}_{mkl} - \mathbf{S}\, T^i{}_{mr} R^r{}_{okl} = \mathbf{S}\, \nabla_m S^i{}_{kl} + \mathbf{S} S^i{}_{rm} S^r{}_{kl}$$

where **S** denotes the sum of terms obtained on permuting cyclically the indices (k, l, m), the coefficients of the terms in $\alpha^k \wedge \alpha^l \wedge \theta^m$ and in $\alpha^k \wedge \theta^l \wedge \theta^m$ vanish in (10.1). Similarly on identifying in (10.2) the terms in $\alpha^k \wedge \alpha^l \wedge \alpha^m$,
$\alpha^k \wedge \alpha^l \wedge \theta^m$, $\alpha^k \wedge \theta^l \wedge \theta^m$, $\theta^k \wedge \theta^l \wedge \theta^m$ we obtain successively

(10.4) $\mathbf{S}\,\nabla_m R^i{}_{jkl} + \mathbf{S}\, S^r{}_{kl} R^i{}_{jrm} - \mathbf{S} P^i{}_{jmr}\, R^r{}_{okl} = 0$

(10.5) $\nabla_{\dot{m}} R^i{}_{jkl} + T^r{}_{km} R^i{}_{jrl} + T^r{}_{lm}\, R^i{}_{jkr} + \nabla_k P^i{}_{jlm} - \nabla_l P^i{}_{jkm}$
$+ S^r{}_{kl} P^i{}_{jrm} - (P^i{}_{jkr} P^r{}_{olm} - P^i{}_{jlr} P^r{}_{okm}) + Q^i{}_{jrm} R^r{}_{okl} = 0$

(10.6) $\nabla_{\dot{m}} P^i{}_{jkl} - \nabla_{\dot{l}} P^i{}_{jkm} + \nabla_k Q^i{}_{jlm} + P^i{}_{jrl}\, T^r{}_{km} - P^i{}_{jrm} T^r{}_{kl}$
$+ Q^i{}_{jrm}\, P^r{}_{okl} - Q^i{}_{jrl}\, P^r{}_{okm} - P^i{}_{jkr}\, Q^r{}_{olm} = 0$

(10.7) $\mathbf{S}\,\nabla_{\dot{m}} Q^i{}_{jkl} + \mathbf{S}\, Q^i{}_{jrm}\, Q^r{}_{okl} = 0$

11. Torsion and curvature defined by a covariant derivation.

Let us suppose that the general linear connection is regular. We denote by ∇ the corresponding covariant derivation in the fibre bundle $p^{-1}TM \to V(M)$. Let $\hat{X}$, $\hat{Y}$, $\hat{Z}$ be three vector fields on V(M) over X, Y, Z belonging to T_{pz}. Then we can express the **torsion** and the **curvature** of the connection by the following formulas:

(11.1) $\tau(\hat{X}, \hat{Y}) = \nabla_{\hat{X}} Y - \nabla_{\hat{Y}} X - \rho[\hat{X}, \hat{Y}]$

(11.2) $\Omega(\hat{X}, \hat{Y})Z = \nabla_{\hat{X}} \nabla_{\hat{Y}} Z - \nabla_{\hat{Y}} \nabla_{\hat{X}} Z - \nabla_{[\hat{X},\hat{Y}]} Z$

τ and Ω being respectively the torsion and the curvature of ∇. They determine two torsion tensors, denoted by S and T, and three curvature tensors R, P, and Q. following the decomposition of vector fields into horizontal and vertical parts.

(11.3) $\tau(H\hat{X}, H\hat{Y}) = S(X, Y)$, $\tau(V\hat{X}, H\hat{Y}) = T(\dot{X}, Y)$

(11.4) $\Omega(H\hat{X}, H\hat{Y}) = R(X, Y)$, $\Omega(H\hat{X}, V\hat{Y}) = P(X, \dot{Y})$,

$$\Omega(V\hat{X}, V\hat{Y}) = Q(\dot{X}, \dot{Y})$$

where $\dot{X} = \mu(V\hat{X})$, $\dot{Y} = \mu(V\hat{Y})$ On deriving the equations (11.1) and (11.2) and on using the Jacobi identity for the three vector fields $\hat{X}$, $\hat{Y}$, $\hat{Z}$ on V(M) we obtain

(11.5) $\mathbf{S}\,\Omega(\hat{X}, \hat{Y})Z = \mathbf{S}\,\nabla_{\hat{Z}}\, S(\hat{X}, \hat{Y}) + \mathbf{S}\,\tau(\hat{Z}, [\hat{X}\,\hat{Y}])$

(11.6) $\mathbf{S}\,\Omega(\hat{X}, \hat{Y})Z + \mathbf{S}\,\Omega(\hat{Z}, [\hat{X}\,\hat{Y}]) = 0$

where **S** denotes the sum of the terms obtained by permuting cyclically $\hat{X}$, $\hat{Y}$, and $\hat{Z}$. The equations (11.5) and (11.6) are called **Bianchi identities.** On decomposing the vector fields into horizontal and vertical components of (11.5), we obtain an identity (see (10.3), and of (11.6) we obtain four identities (see (10.4), (10.5), (10.6) and (10.7)).

CHAPTER II

FINSLERIAN MANIFOLDS

(Abstract) A metric manifold is defined by the data of a tensor field g_{ij} in the restricted sense of degree zero on W(M). To this tensor field is associated a scalar of degree two in the restricted sense $F^2 = g_{ij}(x, v) v^i v^j$ where $F > 0$ is by definition the length of tangent vectors v to M at $x \in M$. With the help of g we can define the scalar product of two vectors of $T_{\pi y}$, $y \in W(M)$, consequently an orthonormal frame on $T_{\pi y}$. Let us denote by E(W, g) the principal fibre bundle on W(M) of orthonormal frames. An infinitesimal connection on E(W, g) is called a Euclidean connection. We give the necessary and sufficient conditions in order that a linear connection on W(M) is naturally associated to a Euclidean connection of directions.

We say that the datum of a function $F > 0$ homogeneous of degree one on V(M) defines a Finslerian metric if it leads to a regular problem of the calculus of variations. The following result is the fundamental theorem of Finslerian Geometry:

Given a Finslerian manifold there exists a regular Euclidean connection such that its torsion tensor S vanishes and the tensor T satisfies a condition of symmetry.

Such a characterization of the Finslerian connection leads us naturally to Cartan's Euclidean connection ([1c], [13]). Using the results of chapter I we establish the fundamental formulas of Finslerian geometry the three curvature tensors, five Bianchi identities are completely made explicit[§ 8]. §9 is devoted to the semi-metric connection and we give a characterization of Berwald connection. We show that there exists an infinity of torsion-free connections of directions attached to F that define the same splitting of the tangent bundle as the Finslerian connection. These connections, differ from Berwald or Cartan connections by a homogeneous tensor t^i_{jk} of degree zero satisfying $t^i_{oj} = 0 = t^i_{ko}$., and have the same flag curvature as Berwald and Cartan connections.

1. Metric Manifolds. Let g_{ij} be a tensor field in the restricted sense of degree zero, symmetric and positive definite. To the tensor field g_{ij} we can associate a scalar in the restricted sense of degree 2 such that

(1.1) $$2L = F^2 = g_{ij}(x, v)\, v^i v^j \qquad (i, j = 1,\ldots n)$$

where F (> 0) is, by definition, the length of the vector v tangent to M at x. Then we say that the data of g makes M *a metric manifold*, g_{ij} is called the metric tensor, L the fundamental function of the metric manifold. With the help of the tensor g we can put a norm on the tangent vector space $T_{\pi y}$ ($x = \pi y$). An orthonormal frame at $y \in W$ is by definition an orthonormal base of the Euclidean vector space $T_{\pi y}$; it is thus an ordered set of n unitary vectors $(e_1, e_2, \ldots e_n)$ of $T_{\pi y}$ such that their scalar products two by two are given by

$$e_i e_j = \delta_{ij}, \qquad (i, j = 1,2,\ldots n),$$

where δ_{ij} is the Kronecker symbol. We denote by E(W, g) the principal fibre bundle of orthonormal frames over W. This space admits a fibre and a structural group identical to the orthogonal group O(n). Let us consider a covering of M by neighbourhoods (U) endowed with orthonormal frames ; let R_U^y an orthonormal frame at $y \in \pi^{-1}(U)$ if $y \in \pi^{-1}(U \cap V)$ we have

$$R_V^y = R_U^y\, C_V^U\, (y),$$

where the matrix C is an element of the group O(n) ; with respect to such a covering the metric of the space becomes

(1.2) $$ds^2 = 2L(x, dx) = \sum_i^n (\alpha^i)^2$$

2. **Euclidean Connections**. An infinitesimal connection on E(W, g) is called a Euclidean connection of directions. Let us consider a covering of M by neighbourhoods endowed with local sections of E(W, g); a Euclidean connection of directions is defined here by the data in each U of a 1-form ω_U^y ($y \in \pi^{-1}(U)$) over $\pi^{-1}(U)$ with values in the Lie algebra of the orthogonal group

O(n). Such a form is represented by an (nxn) skew-symmetric matrix which we denote again by

$$\omega_U^y = (\omega_{ij}), \qquad (\omega_{ij} + \omega_{ji} = 0)$$

its elements are *differential 1-forms of y*. If $y \in \pi^{-1}(U \cap V)$ we have

(2.1) $$R_V^y = R_U^y \, C_V^U(y), \quad y \in \pi^{-1}(U \cap V)$$

where ω_U^y must satisfy the conditions of coherence.

(2.2) $$\omega_V^y = (C_V^U(y))^{-1} \, \omega_U^y \, C_V^U(y) + (C_V^U(y))^{-1} d\,C_V^U(y), \; C \in O(n)$$

We now show that to every Euclidean connection of directions is naturally associated a linear connection. Let us consider the connection defined with respect to a covering of M, endowed with orthonormal frames by the matrices π_U^y such that

$$\pi_U^y = \omega_U^y$$

The linear connection thus defined is evidently independent of the covering chosen. It is called the linear connection associated to the Euclidean connection the absolute differential of the tensor $g_{ij}(y)$ in this connection is given by

(2.3) $$\nabla g_{ij} = dg_{ij} - \omega^h{}_i \, g_{hj} - \omega^h{}_j \, g_{ih}$$

The frames being orthonormal we have $g_{ij} = \delta_{ij}$
The relation (2.3) then becomes

(2.4) $$\nabla g_{ij} = - (\omega_{ji} + \omega_{ij}) = 0$$

It thus follows that the absolute differential of the metric tensor is zero. Conversely let π be a linear connection such that the absolute differential of the tensor metric in this connection is zero ; by (2.3) and (2.4), it is clear that for a covering of M by the neighborhoods endowed with local sections of E(W, g) the matrices of the linear connection are skew-symmetric; on the other hand if

$y \in \pi^{-1}(U \cap V)$ we have

$$R_V^y = R_U^y \, C_V^U \,(y),$$

where C_V^U is an element of the group O(n) ; for the linear connection envisaged we have the condition of coherence

$$\pi_V^y = (C_V^U)^{-1} \, \pi_U^y \, C_V^U + (C_V^U)^{-1} d C_V^U,$$

It thus follows that the matrices $\omega_U^y = \pi_U^y$ define a 1-form of connection with values in the Lie algebra of O(n), that is to say a Euclidean connection. Thus we have the

Theorem. *In order that a linear connection on W be naturally associated to a Euclidean connection of directions it is necessary and sufficient that absolute differential of the tensor metric $g_{ij}(y)$ of the Euclidean connection be zero.*

The Euclidean connection is called *regular* if the associated linear connection is regular. This will be identified with the Euclidean connection in question.

Remarks on the Curvature.---Let ω be a regular linear connection. Let us consider a covering of M by the neighbourhoods endowed with orthonormal frames ; we can therefore put all the indices in lower positions and the curvature form is written

$$\Omega_{ij} = d\omega_{ij} + \omega_{ir} \wedge \omega_{rj} \qquad (\Omega_{ij} = g_{ih}\, \Omega^h{}_j)$$

With respect to the orthonormal frames the ω_{ij} are skew-symmetric; hence

$$\Omega_{ij} + \Omega_{ji} = 0$$

This relation, valid in orthonormal frames, is also valid in arbitrary frames. So it follows that the three curvature tensors of a regular Euclidean connection are skew-symmetric with respect to the first two indices.

3. The System of Generators on W. --- a. To the canonical vector field one can make correspond the vector

$$l = F^{-1}v$$

It is evidently of degree zero, therefore it defines canonically a unitary vector field of degree 0, that is to say a vector field over W. Let ω be a regular Euclidean connection on W, at a point $x \in M$, if the vector tangent space to $p^{-1}(x)$ is referred to the co-frame (μ^i), the vector space tangent to the $\pi^{-1}(x)$ is defined by the equation

$$l_i\, \mu^i = 0 \tag{3.1}$$

The absolute differential of l in the regular Euclidean connection is

$$\beta^i = \nabla l^i = F^{-1}(\theta^i - l^i\, dF), \tag{3.2}$$

l being of unit length we have

$$l_i\, \beta^i = F^{-1}\, (l_i\theta^i - dF) = 0 \tag{3.3}$$

whence

$$\partial_h F = 0, \quad \dot{\partial}_h F = l_h. \tag{3.4}$$

Thus (3.2) becomes

(3.5) $$\beta^i = F^{-1}(\theta^i - l^i l_h \theta^h), \nabla_k l^i = 0,$$

(3.6) $$\nabla_{\dot{k}} l^i = F^{-1}(\delta^i_k - l^i l_k)$$

By (3.5) it is clear that the β^i define on V(M) a 1-form with vector values. This 1-form being of degree zero and satisfying (3.1) can be identified to a 1-form on W with vector values. If we denote by γ^i the restrictions of the β^i to the fibre $\pi^{-1}(x)$, they satisfy

(3.7) $$l_i \gamma^i = 0$$

Thus every linear combination of the γ^i is a 1-form with values in the vector space tangent to $\pi^{-1}(x)$ at y. Conversely the γ^i form a system of generators for these 1-forms , in other words among the γ^i there does not exist any non trivial relation distinct from (3.7). In fact, let

(3.8) $$a_i \gamma^i = 0$$

where the a_i do not all vanish; so in virtue of (3.5) it becomes

$$a_i(\mu^i - l^i l_h \mu^h) = 0,$$

let

$$(a_i - a_h l^h l_i)\, \mu^i = 0$$

the Euclidean connection being regular the μ^i are thus linearly independent; we have

$$a_i = (a_h l^h) l_i$$

and the relation (3.8) is not different from (3.7). It thus follows that *for a regular Euclidean connection the set (α^i, β^i) forms a system of generators for the 1-forms of the vector space tangent to W(M).*

b. Let π be a linear form on W, its inverse image on V(M) by $\eta : V \to W$, which we continue to denote by π, can be put in a unique way under the form

(3.9) $$\pi = C_i\alpha^i + d_i\theta^i$$

where the C_i and d_i are restricted vectors of degree 0 and -1 respectively and the d_i satisfy the condition of homogeneity (chap. I §6)

(3.10) $$d_o = 0$$

If we refer π to the system of generators (α^i, β^i) it becomes

(3.11) $$\pi = a_i\,\alpha^i + b_i\,\beta^i.$$

On substituting in (3.9) the expression for θ^i drawn from (3.5), taking into account (3.10) we obtain

(3.12) $$\pi = C_i\alpha^i + F\,d_i\,\beta^i.$$

On identifying (3.11) to (3.12), we have

$$a_i = C_i, \qquad b_i = Fd_i.$$

Thus referred to the system of generators (α^i, β^i) every 1-form on W can be written under the form (3.11) where the a_i and b_i are restricted vectors of degree 0 and the b_i satisfy

(3.13) $$b_o = 0,$$

Let us prove the uniqueness of the expression for π defined by (3.11). In fact, suppose that $\pi = 0$. The restriction of the fibre $\pi^{-1}(x)$ gives, on taking into account the fact that $b_o = 0$,

$$b_i\,\gamma = F^{-1}\,b_i(\mu^i - l^i l_h\mu^h) = = F^{-1}\,b_i\mu^i = 0.$$

The μ^i are linearly independent , whence

$$b_i = 0.$$

From the fact that $\pi = a_i\,\alpha^i, = 0$ we have $a_i = 0$. This proves the uniqueness. If we relate the tangent vector space to W to the system of generators (α^i, β^i) the 1-form of the regular Euclidean connection of directions is written in a unique way under the form

$$\omega^i_j = \gamma^i_{jk}\alpha^k + B^i_{jk}\beta^k \tag{3.14}$$

where the B^i_{jk} satsfy

$$B^i_{jo} = 0. \tag{3.15}$$

The γ^i_{jk} and B^i_{jk} are restricted quantities of degree zero. On substituting in (3.14) the expression for β^i defined by (3.5) we then have with the notations of chapter I (§5)

$$B^i_{jk}(x, v) = F(x, v)C^i_{jk}(x, v) \tag{3.16}$$

4. Special Connections [1c]. Let ω be a regular Euclidean connection of directions ; the absolute differential of the metric tensor in this connection is zero

$$dg_{ij} = \omega^h_i g_{hj} + \omega^h_j g_{ih} \tag{4.1}$$

We put

$$\gamma_{jik} = g_{jh}\gamma^h_{ik}, \qquad C_{jik} = g_{jh}C^h_{ik} \tag{4.2}$$

Thus (4.1) becomes

$$dg_{ij} = (\gamma_{ijk} + \gamma_{jik})\alpha^k + (C_{ijk} + C_{jik})\theta^k$$

whence

$$\gamma_{ijk} + \gamma_{jik} = \partial_k g_{ij} \tag{4.3}$$

$$C_{ijk} + C_{jik} = \dot{\partial}_k\, g_{ij} \tag{4.4}$$

where ∂_k and $\dot{\partial}_k$ are defined by the formulas 6.8 and 6.9, Chap I
In virtue of (6.6) of the chapter I the $\dot{\partial}_k\, g_{ij}$ satisfy

$$\dot{\partial}_0 g_{ij} = 0$$

Let us suppose that the torsion tensors, associated to ω, satisfy

(4.5) $$S_{ijk} = 0 \qquad (S_{ijk} = g_{ir}\, S^r_{\ jk}),$$

(4.6) $$T_{ijk} = T_{jik}, \qquad (T_{ijk} = g_{ir} T^r_{\ jk})$$

From the formulas (7.4) and (7.5) of chapter I, it follows that there exist, in addition, the following relations among the coefficients of the Euclidean connection in question:

(4.7) $$\gamma_{ijk} - \gamma_{ikj} = b_{ijk} \qquad (b_{ijk} = g_{ir} b^r_{\ jk})$$

(4.8) $$C_{ijk} - C_{jik} = a_{ijk} - a_{jik} \quad (a_{ijk} = g_{ir}\, a^r_{\ jk})$$

On adding the equations (4.4) to (4.8), we get

(4.9) $$C_{ijk} = \frac{1}{2}\,\dot{\partial}_k g_{ij} + \frac{1}{2}(a_{ijk} - a_{jik})$$

On the other hand from the relations (4.3) and (4.7) it follows

(4.10) $$\gamma_{jik} + \gamma_{ikj} = \partial_k\, g_{ij} - b_{ijk}$$

On permuting cyclically the indices i, j, k we obtain

(4.11) $$\gamma_{kji} + \gamma_{jik} = \partial_i g_{jk} - b_{jki}\,,$$

(4.12) $$\gamma_{ikj} + \gamma_{kji} = \partial_j\, g_{ki} - b_{kij}\,.$$

On redoing (4.12) the sum of (4.10) and (4.11) we have, after a change of indices,

(4.13) $$\gamma_{ijk} = \frac{1}{2}(\partial_k g_{ij} + \partial_j g_{ik} - \partial_i g_{kj}) - \frac{1}{2}(b_{ikj} + b_{jik} - b_{kji})$$

Conversely, it is easy to verify that the quantities γ_{ijk} and C_{ijk} defined by (4.9) and (4.13)) satisfy the equations (4.3), (4.4) (4.7) and (4.8), whence

Definition.[1c] *We call a special connection any regular Euclidean connection such that the corresponding torsion tensors satisfy (4.5) and (4.6). The coefficients of such a connection can be expressed by the formulas (4.9) and (4.13).*

5. Case of Orthonormal Frames and Local Coordinates for the Class of Special Connections [1c].

a. Let us consider a covering of M by neighborhoods endowed with orthonormal frames and let (e_i) be such a frame, we then have

$$e_i e_j = \delta_{ij} = g_{ij}$$

in this case, the formulas (4.9) and (4.13) become

(5.1) $$C_{ijk} = \frac{1}{2}(a_{ijk} - a_{jik}) \qquad (C_{ijo} = 0)$$

(5.2) $$\gamma_{ijk} = -\frac{1}{2}(b_{jik} + b_{ikj} - b_{kji})$$

b. If M is covered by neighborhoods (U) endowed with natural frames of local coordinates the quantities a and b vanish, the coefficients of the two kind of the special connection become

(5.3) $$T_{ijk} = \frac{1}{2}\dot{\partial}_k g_{ij} \qquad (T_{ijo} = 0)$$

(5.4) $$\overset{*}{\Gamma}_{ijk} = \frac{1}{2}(\partial_k g_{ij} + \partial_j g_{ik} - \partial_i g_{kj})$$

where ∂_k and $\dot{\partial}_k$ are the Pfaffian derivatives with respect to the local covering $(dx^k, \theta^k = \nabla v^k)$

6.Finslerian Manifolds.

a. In paragraph 1 we have defined the length of the canonical vector v_x tangent to M at x by a positive restricted function $F(x, v_x)$. On substituting dx for v in F and putting

$$ds = F(x, dx), \tag{6.1}$$

F is by definition the arc element. Let $l(x_0,x_1)$ a path in M with origin at x_0 and the extremity at x_1 ; in virtue of (6.1) the length $l(x_0, x_1)$ is defined by

$$s(x_0, x_1) = \int_{l(x_0,x_1)} F(x, \dot{x})\, du \qquad (\dot{x} = \frac{dx}{du}) \tag{6.2}$$

F being restricted of degree 1 the integral of the right hand side is independent of the parametric representation chosen for the path $l(x_0, x_1)$. The first variation of this integral with non fixed extremities $l(x_0, x_1)$ is given by

$$s = \pi_1 - \pi_0 - \int_{l(x_0,x_1)} \left(\frac{d}{du}\frac{\delta F}{\delta \dot{x}^i} - \frac{\delta F}{\delta x^i}\right) \delta x^i\, du \tag{6.3}$$

where π is a restricted linear differential form of degree 0 defined by

$$\pi = \frac{\delta F}{\delta \dot{x}^i}\, \delta x^i \tag{6.4}$$

and π_0 and π_1 correspond to the points x_0 and x_1. We call the *extremal* of the problem of calculus of variations attached to $F(x, \dot{x})$. It is a solution of the differential system of the second order.

(6.5) $$\frac{d}{du}\frac{\delta F}{\delta \dot{x}^i} - \frac{\delta F}{\delta x^i} = 0 \qquad \dot{x} = dx/du, \quad (i = 1, 2, ...n)$$

On choosing for the parametric representation of the path $l(x_0, x_1)$ the length of the arc $s(x_0, x_1) = s$ the system (6.5) becomes

(6.5)' $$\frac{d}{ds}\frac{\delta F}{\delta \dot{x}^i} - \frac{\delta F}{\delta x^i} = 0 \qquad (i = 1, 2, ...n), (\dot{x} = dx/ds)$$

The system admits the first integral

$$F(x, \dot{x}) = 1 \qquad (\dot{x} = dx/ds)$$

It thus follows, on multiplying the two sides of (6.5)' by F, we obtain the differential system of the second order

$$\frac{d}{ds}\frac{\delta L}{\delta \dot{x}^i} - \frac{\delta L}{\delta x^i} = 0 \qquad (i = 1, 2, ...n).$$

where we have put $F^2(x, v) = 2L(x, v)$. Let, on developing,

(6.6) $$\delta^{\cdot\cdot}_{ij} L\, \ddot{x}^j + \delta^{\cdot}_{ij} L \dot{x}^j - \delta_i L = 0 \quad (i, j = 1, 2, ...n).$$

The problem of the calculus of variations attached to the function $F(x, \dot{x})$ is called regular if the coefficient of $\ddot{x}$ in (6.6) is a non degenerate quadratic form., that is to say if

(6.7) $$\det(\delta_{ij} L) \neq 0$$

This problem is called positively regular if $\delta_{ij} L$ is a positive definite quadratic form. It is clear that for a covering of M by neighborhoods endowed with natural frames of local coordinates the $\delta_{ij} L$ are the components of symmetric tensor of order 2 restricted of degree 0 ; we put

(6.8) $$g_{ij}(x, v_x) = \delta_{ij} L(x, v_x).$$

This tensor defines on M the structure of a metric manifold. Let us multiply the two sides of (6.8) by v^i and v^j successively; in virtue of the homogeneity of F, we have

$$g_{ij}(x, v_x)\, v^i v^j = \frac{1}{2} v^j \,\dot{\delta}_j F^2 = 2L = F^2(x, v_x).$$

We are thus led to the following definition :

Definition . *Let M be a differentiable manifold and V(M) the space of non-zero tangent vectors to M. The structure of a Finslerian manifold on M is defined by the data of a function F(x, v_x), positive, positively homogeneous of degree 1 on V(M) leading to a regular problem of the calculus of variations.*

In the following we suppose that F leads to a positively regular problem.

b. In the previous sections we have seen that a structure of a Finslerian manifold determines on M a structure of a metric manifold by the tensor g_{ij} defined by (6.8) ; we propose here to study the converse problem.

Let us consider a metric structure defined by the data of tensor field $g_{ij}(x, v)$ symmetric, restricted of degree zero, and definite positive. To this tensorial field we associate a scalar $2L = g_{ij} v^i v^j$ defining the square of the norm of the tangent vector to M at x. Under what condition is it the metric tensor of a Finslerian manifold ? In other words, under what condition g_{ij} can be deduced from ($F^2 = 2L$) by

(6.9) $$g_{ij} = \ddot{\delta}_{ij} L$$

(6.10) $$2L = g_{ij}(x, v) v^i v^j?$$

It is clear that for a covering of M by neighborhoods endowed with natural frames of local coordinates the right hand side of (6.9) is a tensor if for such a covering g_{ij} is defined by (6.9) we obtain by derivation

$$\delta^{\cdot}_k g_{ij} = \delta_{ijk} L$$

whence

$$\delta^{\cdot}_k g_{ij} = \delta^{\cdot}_i g_{jk} = \delta^{\cdot}_j g_{ki},$$

It thus follows

$$\nu^i \delta^{\cdot}_k g_{ij} = 0. \tag{6.11}$$

Conversely, this condition is sufficient : in fact, if the relation (6.11) is satisfied, from the expression 2L we deduce by derivation

$$\delta^{\cdot}_k L = \frac{1}{2} (\delta^{\cdot}_k g_{ij}) \nu^i \nu^j + g_{kj} \nu^j = g_{kj} \nu^j$$

A second derivation gives us

$$\delta^{\cdot\cdot}_{kl} L = (\delta^{\cdot}_l g_{kj}) \nu^j + g_{kl} = g_{kl}$$

Theorem[1c]. - *In order that the metric tensor* g_{ij} *be the metric tensor of a Finslerian manifold it is necessary and sufficient that we have (6.11.)*

7. Finslerian Connection[1c] The above analysis helps us establish that there exists for the structure of a Finslerian manifold one and only one special linear connection corresponding to the Finslerian metric. The terminology Finslerian connection was introduced in [1c] in analogy with the Riemannian connection. Let g_{ij} be the metric tensor of a Finslerian manifold defined by (6.8) Let ω^i_j be the matrix of the special Euclidean connection associated to g_{ij}.

$$\omega^i_j = \Gamma^i_{jk} dx^k + C^i_{jk} d\nu^k = \overset{*}{\Gamma}{}^i_{jk} dx^k + T^i_{jk} \theta^k \tag{7.1}$$

where the $T^{i}{}_{jk}$ and $\overset{*}{\Gamma}{}^{i}{}_{jk}$ are defined by (5.3) and (5.4). In virtue of (5.3) and (6.9) of the chapter I we have in local coordinates

$$(7.2) \qquad T_{ijk} = \frac{1}{2}\, \partial^{\bullet}_{k}\, g_{ij} = \frac{1}{2}(\delta^{\cdot}{}_{k}\, g_{ij} - T^{r}{}_{ok}\delta^{\cdot}{}_{r}\, g_{ij})$$

The metric tensor g_{ij} of the Finslerian manifold satisfies (6.11); multiplying the two sides of (7.2) by v^{i} we obtain

$$v^{i}\, T_{ijk} = \frac{1}{2}\, v^{i}\, \delta^{\cdot}{}_{k}\, g_{ij} = 0$$

T being symmetric with respect the first two indices , we have

$$(7.3) \qquad T^{i}{}_{ok} = 0.$$

Conversely if the relation(7.3) satisfies after (6.9) of chapter I, the Pfaffian derivatives of the two types $\partial^{\bullet}_{k}$, $\delta^{\bullet}_{k}$(local coordinates) coincide and on multiplying the two sides of (5.3) by v^{i} we find the relation (6.11). Thus (7.3) is equivalent to (6.11) and the torsion tensor of the special connection associated to g_{ij} becomes

$$(7.4) \qquad T_{ijk} = \frac{1}{2}\,\delta^{\cdot}{}_{k}\, g_{ij} = \tfrac{1}{2}\, \delta^{\cdot\cdot\cdot}{}_{kij}\, L, \quad (T_{ojk} = T_{iok} = T_{ijo} = 0),$$

the T and C being related by the formula (5.14) of Chapter I; due to the homogeneity of T they coincide and we have

$$(7.5) \qquad T^{i}{}_{jk} = C^{i}{}_{jk} = \frac{1}{2}\, g^{ir}\, \delta_{rjk}\, L.$$

We have thus evaluated the torsion tensor with the help of g and its first derivatives with respect to v. Before calculating the coefficients Γ and $\overset{*}{\Gamma}$ we remark that in virtue of the homogeneity of the tensor T, the linear connection (7.1) is necessarily regular,

For the matrix (5.6) of chapter I reduces the identity matrix. By the formula (5.13) of chapter I, we have

$$(7.6) \qquad \overset{*}{\Gamma}{}^{i}{}_{jk} = \Gamma^{i}{}_{jr} - T^{i}{}_{jr}\,\overset{*}{\Gamma}{}^{r}{}_{ok}$$

due to the homogeneity of T we have

$$(7.7) \qquad \overset{*}{\Gamma}{}^{i}{}_{ok} = \Gamma^{i}{}_{ok}$$

On the other hand in virtue of (6.8) of chapter I and (7.4), we have

$$(7.8) \qquad \partial_k g_{ij} = \delta_k g_{ij} - 2\, T_{ijr}\overset{*}{\Gamma}{}^{r}{}_{ok}$$

whence the coefficients $\overset{*}{\Gamma}_{ijk}$, defined by (5.4) become

$$(7.9)\ \overset{*}{\Gamma}_{ijk} = \frac{1}{2}(\delta_k g_{ij} + \delta_j g_{ik} - \delta_i g_{kj}) - v^r(T_{ijs}\overset{*}{\Gamma}{}^{s}{}_{rk} + T_{iks}\overset{*}{\Gamma}{}^{s}{}_{rj} - T_{kjs}\overset{*}{\Gamma}{}^{s}{}_{ri})$$

Let us multiply this relation by v^j, taking into account the homogeneity of the tensor T we have

$$(7.10)\ v^j\,\overset{*}{\Gamma}_{ijk} = \frac{1}{2}(\delta_{ik}\,L + v^j\,\delta_j g_{ik} - \delta^{\cdot}{}_{ki}L) - T_{iks}\,\overset{*}{\Gamma}{}^{s}{}_{rj}\,v^r\,v^j$$

Let us multiply the two sides by v^k:

$$(7.11)\ v^k\,v^j\,\overset{*}{\Gamma}_{ijk} = \frac{1}{2}(\delta^{\cdot}{}_{ik}L + v^j\delta_j g_{ik} - \delta^{\cdot}{}_{ki}L)v^k = v^k\,\delta^{\cdot}{}_{ik}L - \delta_i L.$$

We put

$$(7.12) \qquad 2G^r(x, v) = \overset{*}{\Gamma}{}^{r}{}_{jk}\,(x, v)\,v^j\,v^k$$

whence

$$(7.13) \qquad 2g_{ir}G^r(x, v) = \overset{*}{\Gamma}_{ijk}v^jv^k = v^k\,\delta^{\cdot}{}_{ik}L - \delta_i L.$$

The above relation determines G as a function of L and its derivatives, ; we notice, in addition, that last term of this relation figure in the differential system of extrema (6.6) ; a derivation with respect to v gives us

(7.14) $g_{is}\delta^{\cdot}_{r}G^{s} = \frac{1}{2}(\delta^{\cdot}_{ir}L + v^{k}\delta_{k}g_{ir} - \delta^{\cdot}_{ri}L) - 2T_{irs}G^{s} = v^{j}\overset{*}{\Gamma}_{ijr},$

whence

(7.15) $$\delta^{\cdot}_{k}G^{r} = \overset{*}{\Gamma}{}^{r}_{ok}$$

On substituting (7.15) in (7.9) we obtain
(7.16)

$$\overset{*}{\Gamma}_{ijk} = \frac{1}{2}(\delta_{k}g_{ij} + \delta_{j}g_{ik} - \delta_{i}g_{kj}) - (T_{ijs}\delta^{\cdot}_{k}G^{s} + T_{iks}\delta^{\cdot}_{j}G^{s} - T_{kjs}\,\delta^{\cdot}_{i}G^{s}).$$

To obtain the expression of $\overset{*}{\Gamma}{}^{r}_{jk}$ it suffices to multiply the two sides of (7.16) by g^{ir}.

Thus the coefficients $\overset{*}{\Gamma}{}^{i}_{jk}$ are completely determined by the metric tensor of the Finslerian manifold and its first derivatives. As to the coefficients Γ^{i}_{jk} we have by (7.6),

(7.17) $$\Gamma_{ijk} = \frac{1}{2}(\delta_{k}g_{ij} + \delta_{j}g_{ik} - \delta_{i}g_{kj}) + T_{kjs}\delta^{\cdot}_{i}G^{s} - T_{iks}\,\delta^{\cdot}_{j}G^{s}$$

the Γ and C determine then a unique regular linear connection on W. Conversely it is easy to show that the quantities Γ and C defined by (7.17) and (7.4) determine canonically a special linear connection associated to the tensor g, whence :

Theorem [1c], [13]. *Given a Finslerian manifold M there exists a special Euclidean connection and only one which corresponds to the Finslerian metric. I call this connection the Finslerian connection.*

8. Curvature Tensors of the Finslerian Connection.

a. The formulas of the paragraph 7, chapter I giving the curvature tensors of a general regular linear connection can be applied in particular to the Finslerian connection. We reason in local coordinates. We recall that in virtue of homogeneity of the torsion tensor T of the Finslerian connection the matrix L [(5.8), chap. I], as well as its inverse M, reduce to the identity matrix (local coordinates). After (7.23), (7.26), and (6.8) of chapter I, we then have

$$(8.1)\quad R^i_{jkl} = \overset{*}{R}{}^i_{jkl} + T^i_{jr}\overset{*}{R}{}^r_{okl}$$

$$(8.2)\quad \overset{*}{R}{}^i_{jkl} = (\delta_k \overset{*}{\Gamma}{}^i_{jl} - \dot{\delta}_s \overset{*}{\Gamma}{}^i_{jl}\, \dot{\delta}_k G^s) - (\delta_l \overset{*}{\Gamma}{}^i_{jk} - \dot{\delta}_s \overset{*}{\Gamma}{}^i_{jk}\, \dot{\delta}_l G^s) + (\overset{*}{\Gamma}{}^i_{rk}\overset{*}{\Gamma}{}^r_{jl} - \overset{*}{\Gamma}{}^i_{rl}\overset{*}{\Gamma}{}^r_{jk}).$$

We now establish the following proposition :
The curvature tensor P of the Finslerian connection is expressed in terms of the torsion tensor T and its covariant derivative of the type ∇_h and the curvature tensor Q is deduced algebraically from the tensor T
In fact, after (7.24) and (7.27) of chapter I, we have

$$(8.3)\quad P^i_{jkl} = \overset{*}{P}{}^i_{jkl} + T^i_{jr}\overset{*}{P}{}^r_{okl}$$

$$(8.4)\quad \overset{*}{P}{}^i_{jkl} = \nabla_k T^i_{jl} - \dot{\delta}_l \overset{*}{\Gamma}{}^i_{jk}.$$

On the other hand

$$\dot{\delta}_l (\overset{*}{\Gamma}_{sjk}) = \dot{\delta}_l (g_{si}\overset{*}{\Gamma}{}^i_{jk}) = g_{si}\, \dot{\delta}_l (\overset{*}{\Gamma}{}^i_{jk}) + 2\, T_{sil}\, \overset{*}{\Gamma}{}^i_{jk}\, .$$

On evaluating the left hand of this relation using (7.16), it becomes

(8.5) $$g_{si}\,\delta_l^{\bullet}\,(\overset{*}{\Gamma}{}^i{}_{jk}) = (\delta_k T_{sjl} + \delta_j T_{skl} - \delta_s T_{kjl}) - 2T_{srl}\,\overset{*}{\Gamma}{}^r{}_{jk}$$
$$- (\delta_l^{\bullet} T_{sjr}\,\delta^{\cdot}_k G^r + \delta_l^{\bullet} T_{skr}\,\delta^{\cdot}_j G^r - \delta_l^{\bullet} T_{kjr}\,\delta^{\cdot}_s G^r)$$
$$-(T_{sjr}\,\delta_{kl}^{\bullet\bullet} G^r + T_{skr}\,\delta_{jl} G^r - T_{kjr}\,\delta_{sl} G^r).$$

On multiplying the two sides by v^j we obtain

(8.6) $$v^j\,\delta^{\cdot}_l\overset{*}{\Gamma}{}^i{}_{jk} = \nabla_0 T^i{}_{kl}.,$$

On deriving with respect v^j the relation (7.15) we have

(8.7) $$\delta_{kl}^{\bullet\bullet} G^r = \overset{*}{\Gamma}{}^r{}_{lk} + \nabla_0 T^r{}_{kl}$$

Taking into account this relation , (8.5) becomes

(8.8)
$$\delta^{\cdot}_l\,\overset{*}{\Gamma}{}^i{}_{jk} = \nabla_k T^i{}_{jl} + \nabla_j T^i{}_{kl} - \nabla^i T_{kjl} + T^r{}_{kj}\nabla_0 T^i{}_{rl} - T^i{}_{kr}\nabla_0 T^r{}_{jl} - T^i{}_{jr}\nabla_0 T^r{}_{kl}$$

whence

(8.9) $$P^i{}_{jkl} = \nabla^i T_{kjl} - \nabla_j T^i{}_{kl} + T^i{}_{kr}\,\nabla_0 T^r{}_{jl} - T^r{}_{kj}\,\nabla_0 T^i{}_{rl}.$$

We deduce from it

(8.10) $$P^i{}_{j0l} = P^i{}_{jk0} = 0$$

As to the curvature tensor Q, we have by (7.25) and (7.28) of chapter I

(8.11) $$Q^i{}_{jkl} = \overset{*}{Q}{}^i{}_{jkl} + T^i{}_{jr}\,Q^r{}_{0kl},$$

(8.12) $$\overset{*}{Q}{}^i{}_{jkl} = (\delta_k^{\bullet} T^i{}_{jl} - \delta_l^{\bullet} T^i{}_{jk}) + (T^r{}_{jl}\,T^i{}_{rk} - T^i{}_{rl}\,T^r{}_{jk})$$

On the other hand

$$\delta_{\dot k} T^i_{\ jl} = \delta_{\dot k} (g^{ir} T_{rjl}) = - 2T^i_{\ rk} T^r_{\ jl} + g^{ir} \delta_{\dot k} T_{rjl}$$

On substituting this relation in (8.12), we obtain

(8.13) $$Q^i_{\ jkl} = \overset{*}{Q}{}^i_{\ jkl} = T^i_{\ rl} T^r_{\ jk} - T^i_{\ rk} T^r_{\ jl}$$

It satisfies

(8.14) $$Q_{0jkl} = Q_{iokl} = Q_{ij0l} = Q_{ijk0} = 0.$$

Thus we have established the proposition.

b. In paragraph 10 of chapter I we have defined the complete list of Bianchi identities for a general regular linear connection. These five identities become in the case of a Finslerian connection [1].

(8.15) $$\mathbf{S}\, R^i_{\ mkl} - \mathbf{S}\, T^i_{\ mr} R^r_{\ okl} = 0$$

(8.16) $$\mathbf{S}\, \nabla_m R^i_{\ jkl} - \mathbf{S}\, P^i_{\ jmr} R^r_{\ okl} = 0$$

(8.17) $$\nabla^{\cdot}_m R^i_{\ jkl} + T^r_{\ km} R^i_{\ jrl} + T^r_{\ lm} R^i_{\ jkr} + \nabla_k P^i_{\ jlm} - \nabla_l P^i_{\ jkm} - (P^i_{\ jlr} \nabla_0 T^r_{\ km} - P^i_{\ jkr} \nabla_0 T^r_{\ lm}) + Q^i_{\ jrm} R^r_{\ 0kl} = 0$$

(8.18) $$\nabla^{\cdot}_m P^i_{\ jkl} - \nabla^{\cdot}_l P^i_{\ jkm} + \nabla_k Q^i_{\ jlm} + P^i_{\ jrl} T^r_{\ km} - P^i_{\ jrm} T^r_{\ kl} + Q^i_{\ jrl} \nabla_0 T^r_{\ km} - Q^i_{\ jrm} \nabla_0 T^r_{\ kl} = 0$$

(8.19) $$\mathbf{S}\, \nabla^{\cdot}_m Q^i_{\ jkl} = 0$$

where we have denoted by ∇_m and $\nabla^{\cdot}_m$ the two covariant derivations in the Finslerian connection and by **S** the sum of terms obtained by permuting cyclically the indices (k, l, m). In virtue of (8.1), the identity (8.15) becomes

(8.20) $$\mathbf{S}_{(j,k,l)} \overset{*}{R}_{ijkl} = 0$$

Let us multiply the two sides by v^i

(8.21) $$\mathbf{S}_{(j,k,l)} \overset{*}{R}_{ojkl} = 0$$

Let us multiply this relation by v^k

(8.22) $$\overset{*}{R}_{ojol} = \overset{*}{R}_{oloj}.$$

The tensor R_{ijkl} being skew symmetric with respect to the indices i and j, k and l, we have

(8.23) $$\overset{*}{R}_{ijkl} = - \overset{*}{R}_{jikl} - 2T_{ijr} \overset{*}{R}{}^{r}{}_{okl}\ .$$

(8.24) $$2\, R_{ijkl} = \overset{*}{R}_{ijkl} - \overset{*}{R}_{jilk}$$

Taking into account (8.23), the identity (8.20) becomes

(8.25) $$\overset{*}{R}_{jikl} - \overset{*}{R}_{kijl} = - \overset{*}{R}_{lijk} - 2(T_{ijr} \overset{*}{R}{}^{r}{}_{okl} + T_{kir} \overset{*}{R}{}^{r}{}_{olj} + T_{lir} \overset{*}{R}{}^{r}{}_{ojk})$$

On changing in the above relation the indices j, i, k, l into k, l, j, i respectively, and on adding the relation thus obtained to (8.25), and taking into account (8.24) we have

(8.26) $$(\overset{*}{R}_{jikl} - \overset{*}{R}_{kijl}) + (\overset{*}{R}_{klji} - \overset{*}{R}_{jlki})$$
$$= -2R_{lijk} - 2[T_{ijr} \overset{*}{R}{}^{r}{}_{okl} + T_{kir} \overset{*}{R}{}^{r}{}_{olj} + T_{lkr} \overset{*}{R}{}^{r}{}_{oji} + T_{jlr} \overset{*}{R}{}^{r}{}_{oik}]$$

Similarly, on changing in (8.26) the indices l, i, j, k into j, k, l, i respectively and on putting in (8.26) the relation thus obtained we have

(8.27) $$R_{lijk} - R_{jkli} = T_{ijr} \overset{*}{R}{}^{r}{}_{olk} + T_{kir} \overset{*}{R}{}^{r}{}_{ojl} + T_{lkr} \overset{*}{R}{}^{r}{}_{oij} + T_{jlr} \overset{*}{R}{}^{r}{}_{oki}$$

9. Almost Euclidean Connections

Let c : [0, 1] → M a differentiable path in M. For $t_0, t_1 \in [0, 1]$, the length of the arc joining $x_0 = c(t_0)$ to $x_1 = c(t_1)$ is defined with the help of the function F by

$$(9.1) \qquad s = \int_{t_o}^{t_1} F(x^i \dot{x}^i)dt \qquad (\dot{x}^i = \frac{dx^i}{dt})$$

On choosing the parametric presentation of $C(x_0, x_1)$ the arc length $s = s(x_0, x_1)$, as in §6 the extremal of (9.1) is a solution of the system of a second order differential equation, defined by (6.6) which we write,

$$(9.2) \qquad \frac{d^2x^i}{ds^2} + g^{ir}(\delta_k g_{rj} - \frac{1}{2}\delta_r g_{jk})\, \dot{x}^j \dot{x}^k = 0$$

where g^{ir} is the inverse of g_{rj}. Let us look for a torsion-free connection of directions ϖ such that the geodesic of ϖ coincides with the extremal of (9.1). If the $\widetilde{\Gamma}^i_{rk}$ are the coefficients of this connection, the system of differential equations representing the geodesics of this equation can be put under the form

$$(9.3) \qquad \frac{d^2x^i}{ds^2} + \widetilde{\Gamma}^i_{jk}(x, \dot{x}^j)\, \dot{x}^j \dot{x}^k = 0,$$

If $\widetilde{D}_k$ is the covariant derivation associated to ϖ, from (9.2) and (9.3) it follows:

$$(9.4) \qquad \widetilde{D}_k\, g_{rj}\, v^j v^k = \frac{1}{2}\widetilde{D}_r\, g_{jk} v^j v^k$$

where $\dot{x} = v$. The relation (9.4) determines *a necessary and sufficient condition in order that the geodesic of the connection ϖ of directions without torsion coincides with the extremal of (9.1).*

This being the case, a torsion-free connection of directions is necessarily *regular.* For the map μ becomes the identity map. On the other hand its curvature 2-form on W(M) decomposes into a 2-form of the type (2, 0), that is to say two times horizontal and a 2-form of the type (1,1), one time horizontal and one time vertical, which we denote by $\widetilde{P}$. We then have

Theorem. *Let (M, g) be a Finslerian manifold. There exists a unique torsion-free connection* $\widetilde{D}$ *of directions such that*

1) *the geodesics of* $\widetilde{D}$ *coincide with the extremal of the variational problem corresponding to F.*

2) *the second curvature tensor of this connection satisfies*

(9.5) $$\widetilde{P}(X, \dot{Y})v = 0$$

The coefficients of the connection ϖ *are defined by :*

(9.6) $$\widetilde{\Gamma}^{i}_{jk} = \frac{1}{2} g^{ir}(\partial_k g_{jr} + \partial_j g_{rk} - \partial_r g_{jk}) + \widetilde{D}_o T^{i}_{jk}$$

$$\partial_k = \delta_k - \widetilde{\Gamma}^{r}_{ok}\, \delta^{\bullet}_{r} \qquad \left(\delta_k = \frac{\delta}{\delta x^k},\ \delta^{\bullet}_{r} = \frac{\delta}{\delta v^r}\right)$$

where T is the torsion tensor of the Finslerian connection defined by (7.4).

Proof. On deriving vertically (9.4) and on taking into account (9.5) we obtain

(9.7) $$v^k(\widetilde{D}_k g_{rj} + \widetilde{D}_j g_{rk} - \widetilde{D}_r g_{jk}) = 0$$

(9.8) $$v^k \widetilde{D}_k g_{jr} = 0, \quad v^k \widetilde{D}_j g_{rk} = v^k \widetilde{D}_r g_{jk}$$

Deriving once again vertically the relation (9.8), in virtue of (9.5) we get

(9.9) $$\widetilde{D}_l g_{ij} + \widetilde{D}_o(\partial^{\bullet}_l g_{ij}) = 0$$

Thus $\widetilde{D}_l g_{ij}$ is completely symmetric. From the fact that $\widetilde{D}$ is torsion-free from (9.9), we obtain, on permuting cyclically the indices i, j, and k, the three relations on subtracting the third one from the sum of the other two, we obtain the formula (9.6). *This is called the Berwald connection.* We denote by H and G the corresponding curvature tensors.

It is clear that if g is Riemannian, the condition (9.5) is automatically satisfied, and $\widetilde{D}$ is none other than the Riemannian connection.

The connection ϖ defined by (9.6) can be obtained ([8]) on putting in $2G^i$ the homogeneous expression of second degree in $\dot{x}$ in the differential equation (9.2) and on deriving twice vertically $\delta_{\dot{j}} G^i = G^i_j$; $\delta_{\dot{j}\dot{k}} G^i = G^i_{jk} = \widetilde{\Gamma}^i_{jk}$. [this method was suggested to Berwald by Emmy Noether (see E. Cartan Oeuvres Complètes [14] p. 1393)]. However, if we abandon the hypothesis (9.5) $\widetilde{P}(X, \dot{Y})v \neq 0$ and we suppose $\widetilde{P}(v, \dot{Y})v = 0$, we still have the relation (9.8). In this case G^i_{jk} is related to $\widetilde{\Gamma}^i_{jk}$ by $G^i_{jk} = \widetilde{\Gamma}^i_{jk} + t^i_{jk}$ where t^i_{jk} is a symmetric tensor, homogeneous of degree zero in v and satisfies $t^i_{ok} = t^i_{ko} = 0$. The choice of this tensor determines the connection $\widetilde{\Gamma}$. We note that this connection defines the same splitting of the tangent bundle as the Berwald connection and that the expression of the curvature $\widetilde{H}^i_{ojo}$ of this connection, called flag curvature, is identical to that of Berwald and Cartan. In the following we denote by ∇ and D respectively the connections of Finsler and Berwald ($\widetilde{D} = D$) associated to g and we put :

$$\nabla v^i = Dv^i = \theta^i$$

Remark. D. Bao and Z. Shen in their work on Finslerian Geometry use constantly a semi-metric connection which they call Chern connection. Now S. S. Chern tried to obtain by the method of local equivalence a set of semi-Euclidean connections for the Finslerian metric (E. Engel Zur Flächen theorie. I Leopz Ber. 1901pp 4004-412, E. CARTAN sur un problème d'équivalence et la théorie des espaces métriques généralisées, Mathematica t.4 1930pp 114-136). He establishes the following theorem

Theorem. There is a unique set of local 1-forms ω^i_j on TM/{0} such that

$$d\omega^i = \omega^j \wedge \omega^i_j$$
$$dg_{ij} = g_{kj}\,\omega^k_i + g_{ik}\,\omega^k_j + 2\,C_{ijk}\,\omega^{n+k}$$
$$\omega^{n+i} = dy^i + y^j\,\omega^i_j$$

Unfortunately, the expressions defining ω^i_j are none other than the horizontal part of the Cartan connection denoted by

$$\overset{*}{\Gamma}{}^{i}_{jk}\ (x, v)\, dx^k$$

(see E. Cartan C.R. Académie .Sc.t 198 (1933) pp 582-586. and 'Les espaces de Finsler Paris Hermann (1934).

CHAPTER III

ISOMETRIES AND AFFINE VECTOR FIELDS ON THE UNITARY TANGENT FIBRE BUNDLE

(Abstract). This chapter is devoted to the study of infinitesimal isometries of a compact Finslerian manifold without boundary, and affine infinitesimal transformation of a regular linear connection of directions.([1], [1a], [1b]). We recall the calculus rules of Lie derivatives of a tensor field in the large sense and of a form of a regular linear connection of vectors. Let L be the Lie algebra of infinitesimal transformations of M. To an $X \in L$ is associated a certain endomorphism A_X of T_{pz} whose expression contains the torsion tensor T. Let $\underline{A}_z(L)$ be the Lie algebra of endomorphisms of T_{pz} corresponding to the elements of L. We establish then a relation between $A_{[X,Y]}$, A_X, A_Y and the curvature of the linear connection, generalizing from the Riemannian case due to B. Kostant ([26], [1]). For the study of compact Finslerian manifolds we establish the divergence formulas for the horizontal 1-forms and for the vertical 1-forms on W(M) ([1],[2]). Next we study the 1-parameter group of infinitesimal transformations that leave invariant the splitting of the tangent bundle defined by Finslerian connection. We give a local characterization of isometries. In case M is compact and without boundary we prove the largest connected group of transformations that leave invariant the splitting defined by the Finslerian connection coincides with the largest group of isometries. We establish a formula linking the square of the vertical part of the lift of an isometry X on V(M) and the integral involving an expression of the flag curvature (R(X,u)u,X). If this form is negative definite the isometry group is finite. Finally, to every infinitesimal isometry X of a Finslerian manifold is associated an anti-symmetric endomorphism whose square of the module modulo a divergence puts in evidence a quadratic form φ depending on two Ricci tensor R_{ij} and P_{ij}.[1b]. We determine the conditions on them so that the isometry group of the manifold is finite. We study the particular case of $P_{ij} = 0$
In paragraph §10 we give a characterization of affine infinitesimal transformation (respectively partial) of regular linear connections of vectors. In paragraph § 11 we show that the Lie derivative L(X) commutes with the covariant derivatives of two types ∇ and $\nabla^{\bullet}$ (respectively of type ∇) when X defines an affine infinitesimal transformation (respectively partial), and conversely. Let L be the Lie algebra of affine transformations of a generalized linear connection, and $\widetilde{L}$ its lift on V(M). The Lie algebra $\underline{A}z(L)$, corresponding to L is the Lie algebra of a connected group Kz(L) of linear transformations of Tpz. The study of this group is the objective of paragraphs §

12, 13 and 14. In the case when the Lie algebra $\widetilde{L}$ is transitive on V(M) we have a relation of inclusion

$$\sigma z \subset Kz(L) \subset N_O(\sigma z)$$

where σz is the group of restricted homogeneous holonomy at $z \in V(M)$ and $N_O(\sigma z)$ indicates the passage to the connected normalizer [1].. The group Kz(L) has been introduced by B. Kostant[26] in the Riemannian case. In conclusion we also study the case of affine infinitesimal transformation of a Finslerian connection.

1.Local Group of 1-parameter local transformations and Lie derivative

a A vector field X on M generates a local 1-parameter group of local transformations of M by the integration of the differential system

$$(1.1) \qquad \frac{dx(u)}{du} = X\,x(u) \qquad (u \in \mathbf{R})$$

starting from an initial point x(0) = x. From the existence theorem for the solutions of the differential system it follows that it is possible to find neighbourhoods $U(\overline{x})$ and positive numbers $\varepsilon(\overline{x})$ for every $\overline{x} \in M$ such that (1.1) admits a solution denoted by

$$(1.2) \qquad x(u) = \exp(uX)\,x$$

defined for $|u| < \varepsilon(\overline{x})$ starting from a point $x \in U(\overline{x})$ and satisfying the group relation

$$(1.3) \qquad \exp[(u+u')X]\,x = \exp(uX)\exp(u'X)x$$

provided that the two sides are defined. In general, the numbers $\varepsilon(\overline{x})$ depend on $\overline{x}$; if we can choose $\varepsilon(\overline{x})$ independent of $\overline{x}$ then exp (uX) can be defined for every $x \in M$, and for $|u| < \infty$, in this case X generates a 1-parameter group of global transformations of

M. If the manifold M is compact every vector field on M generates a 1-parameter group of global transformations of M.

b. The differentiable map exp (uX) is defined over U(x) for every $|u| < \varepsilon(\bar{x})$ and it admits a linear tangent map denoted exp (uX)′ which defines an isomorphism of T_x with $T_{x(u)}$. To every element z of V(M) above $U(\bar{x})$ corresponds thus an element z(u). From the map it follows that for each $U(\bar{x})$ and $\varepsilon(\bar{x})$ a differentiable map is defined in $p^{-1}(U(\bar{x}))$

$$(1.4) \qquad \exp(u\hat{X}): z \to z(u)$$

$\exp(u\hat{X})$ defines a local 1-parameter group of local transformations of V(M) which we call the extended group. By its definition we have

$$(1.5) \qquad p.\exp(u\hat{X})\,z = \exp(uX).\,pz,$$

we conclude from it, on deriving with respect to u that the vector field $\hat{X}$ over V(M) is projectable by p

$$(1.6) \qquad p\,\hat{X} = X$$

From (1.5) it follows that

$$(1.7) \qquad p'.\exp(u\hat{X})' = \exp(uX)'.p',$$

$\hat{X}$ is called a lift of X to V(M). Let (α^i, θ^i) be a co-frame at $z \in V(M)$, where the θ^i are supposed arbitrary ; if we denote by i the operator of the interior product, we put

$$(1.8) \qquad i(\hat{X})\,\alpha^i = X^i, \quad i(\hat{X})\,\theta^i = \dot{X}^i$$

Thus at the point $z \in V(M)$ the vector $\hat{X}$ has for components with respect to the frames in question X^i and $\dot{X}^i$; in particular for the local co-frame (dx^i, dv^i) at $z \in V(M)$ the components of $\hat{X}$ are

$$(1.9) \qquad \frac{dx^i(u)}{du}\Big|_{u=0} = X^i, \qquad \frac{dv^i(u)}{du}\Big|_{u=0} = v^j\,\delta_j\, X^i = \delta_0 X^i$$

We remark that in a change of frame the quantities $\delta_0 X^i$ do not transform themselves according to the components of a vector.

c. Let Φ be an r-form on V(M) with values in a vector space N. We call the infinitesimal transformation of Φ by $\hat{X}$ at z the r-form $L(\hat{X})\Phi$ [by abuse of notation $L(X)\Phi$] at z with values in N defined by

$$(1.10) \qquad L(X)\Phi = i(\hat{X})d\Phi + di(\hat{X})\,\Phi$$

It is clear that L(X) is a derivation of degree 0 which commutes with d. If f is a function with real values on V(M) by (1.10) we have

$$L(X)f = i(\hat{X})\,df = X^k\,\partial k\,f + \dot{X}^k\,\partial_{\dot{k}}\,f,$$

where X^k and $\dot{X}^k$ are the components of $\hat{X}$ with respect to the co-frame (α^i, θ^i) ; relative to the co-frame (dx^i, dv^i) this relation can be written

$$L(X)f = X^k\,\delta_k f + \delta_0\,X^k\,\delta_{\dot{k}}\,f,$$

L(X) commutes with the operations δ_k and $\delta_{\dot{k}}$ of ordinary derivations.

Given a tensorial field in the large sense (or a connection form) on V(M), for it to be invariant by the extended group exp $(u\hat{X})$, it is necessary and sufficient that its Lie derivative with respect to $\hat{X}$

is 0. Besides, it is clear that the canonical vector v is invariant by the extended group exp(u $\hat{X}$), in other terms, exp u($\hat{X}$) does not change the differentiable structure of the manifold V(M)

To a frame R^z, attached to z, corresponds, by the action of the group exp (u $\hat{X}$), a frame $R^{z(u)}$ at z(u), the differentiable map

$$\exp (u\hat{X}) : R^z \rightarrow R^{z(u)}$$

defines a local 1-parameter group of local transformations of the fibre bundle $p^{-1}E(M)$ generated by the vector field $\hat{X}$ which is a lift of X. We will define in the same way as in classical differential geometry [28] the infinitesimal transformation of a form on $p^{-1}E(M)$.

Given a form Λ on $p^{-1}E(M)$ with values in a vector space N, the infinitesimal transformation of Λ by $\hat{X}$ is by definition the Lie derivative by X and will be denoted by $L(X)\Lambda$. One proves that [28]

1. If Λ is a tensorial q-form of type R(G) (linear representation of G in a vector space), $L(X)\Lambda$ is also a tensorial q-form of the same type.

2. If ω is a 1-form of connection with values in the Lie algebra GL(n, **R**) of adjoint type, $L(X)\omega$ is tensorial 1-form of adjoint type.

Let Y be a vector field in the large sense, and Λ a semi-basic 1-form on V(M), that is to say it depends only on the 1-form of the base. Since L(X) satisfies the Leibniz rule for the tensor product and commutes with trace operator we have

$$L(X)\, i(Y)\Lambda = i(Y)\, L(X)\Lambda + i\,(L(X)\, Y)\Lambda . \tag{1.11}$$

2.Local Invariant Sections.

a. Let $\hat{X}$ be a vector field on V(M), a point z of V(M) is called ordinary if it is not a zero of $\hat{X}$; z is a zero of the first kind if every neighbourhood $\overline{U}$ of z contains ordinary points ; z is a zero of the second kind if there exists a neighbourhood $\overline{U}$ of z such that every point of $\overline{U}$ is a zero of $\hat{X}$ in this case exp(u$\hat{X}$) defines the identity on $\overline{U}$ and the infinitesimal transformation of a tensor in the large sense or of a connection at the point z is zero. Concerning the zeros of the first kind, if t is a tensorial field in the large sense (connection form); if we evaluate [L(X)t]z at a ordinary point $z \in V(M)$ we can obtain $[L(X)t]z_0$ at a zero z_0 of the first kind by passing to the limit.

b. Given an ordinary point $z \in V(M)$ there exists a neighbourhood $U \subset V(M)$ of z consisting of ordinary points ; if Rz is the frame attached to z the points of $p^{-1}E(M)$ defined by Rz(u) = exp(u$\hat{X}$) Rz belong to the trajectory of $\hat{X}$ starting from Rz. We are thus led to the following definition:

Definition. *Given a neighbourhood $\overline{U} \subset V(M)$ consisting of ordinary points we call a local section μ of $p^{-1}E(M)$ generated by the trajectories of $\hat{X}$ will be called a local invariant section of X over $\overline{U}$.*

On $\overline{U}$ the frames of the local section define a vector field $e_i(z)$, invariant by X; if the n vectors $e_i(z)$ determine the frame at z, $R^z = \mu z$, we have

$$L(X)\, e_i(z) = 0 \tag{2.1}$$

On the other hand, we have the equality between the scalars

$$i(e_j)\, \alpha^i = \delta^i_j$$

The Lie derivative of the right hand side vanishes. So, taking into account (1.11), we have

$$L(X)\, i(e_j)\alpha^i = i(L(X)e_j)\alpha^i + i(e_j)\, L(X)\alpha^i = 0$$

For (2.1) to be satisfied it is necessary and sufficient that

$$i(e_j)\, L(X)\alpha^i = 0$$

whatever be e_j, whence,

$$L(X)\alpha^i = 0 \tag{2.2}$$

c. Let Λ be a form on $p^{-1}E(M)$ with values in a vector space N. Let $\overline{\Lambda}$ the form induced on the local section of $p^{-1}E(M)$ over $\overline{U}$. On $\overline{U}$ there exist a form $\Phi_{\overline{U}}$ with values in N such that

$$\mu^* \overline{\Lambda} = \Phi_{\overline{U}}$$

For the local section in view, we obtain

$$\mu^* L(\hat{X})\overline{\Lambda} = L(X)\, \Phi_{\overline{U}} = [i(\hat{X})\, d + di(\hat{X})]\, \Phi_{\overline{U}} \tag{2.3}$$

This formula helps us evaluate $L(\hat{X})\overline{\Lambda}$. In virtue (2.3), the relation (2.2) becomes

$$[L(X)\, \alpha]^i = L(X)\, \alpha^i = 0 \tag{2.4}$$

3. Introduction of a regular linear connection

a. Let us consider a covering of M by neighborhoods U, let ω be a regular linear connection on V(M) represented in each $p^{-1}(U)$ by the matrices ω^i_j of 1-forms defined by [(5.7), chap I]. Let us introduce the connection ϖ said to be associated to ω defined by the matrices ϖ^i_j such that

$$\varpi^i_j = \omega^i_j + S^i_{jk}\, \alpha^k \tag{3.1}$$

where S is the first torsion tensor of ω. If $\overline{\nabla}$ denotes the absolute differential in the associated connection, from (3.1) it follows

(3.2) $$\bar{\theta}^{i} = \bar{\nabla} v^{i} = \theta^{i} + S^{i}_{0k}\,\alpha^{k}$$

After (3.1) it is clear that the associated connection is regular. Writing down the two connections ω and $\bar{\omega}$ with respect to the co-frames (dx, θ) and $(dx, \bar{\theta})$ respectively, the relation (3.1) becomes

$$\begin{aligned}\bar{\overset{*}{\Gamma}}{}^{i}_{jk}dx^{k} + \bar{T}^{i}_{jk}\,\bar{\theta}^{k} &= \overset{*}{\Gamma}{}^{i}_{jk}dx^{k} + T^{i}_{jk}\theta^{k} - (\overset{*}{\Gamma}{}^{i}_{jk} - \overset{*}{\Gamma}{}^{i}_{kj})dx^{k} \\ &= \overset{*}{\Gamma}{}^{i}_{kj}dx^{k} + T^{i}_{jk}\theta^{k}\end{aligned}$$

where $\bar{\overset{*}{\Gamma}}$ and $\bar{T}$ are the coefficients of the connection $\bar{\omega}$; on substituting in this relation the θ^{k}, defined by (3.2), and on identifying the two sides we obtain

(3.3) $$\bar{\overset{*}{\Gamma}}{}^{i}_{jk} = \overset{*}{\Gamma}{}^{i}_{kj} - T^{i}_{jh}\,S^{h}_{ok}$$

(3.4) $$\bar{T}^{i}_{jk} = T^{i}_{jk}$$

If $\bar{S}$ denotes the torsion tensor of $\bar{\omega}$, from (3.3) we have

(3.5) $$\bar{S}^{i}_{jk} = T^{i}_{jh}\,S^{h}_{ok} - T^{i}_{kh}\,S^{h}_{0j} - S^{i}_{jk}$$

Thus the torsion tensors of $\bar{\omega}$ are defined by (3.4) and (3.5). Let Y be a vector field in the large sense, its absolute differential with respect to $\bar{\omega}$ can be put under the form

(3.6) $$\bar{\nabla} Y^{i} = \bar{\nabla}_{k} Y^{i}\alpha^{k} + \bar{\nabla}_{\dot{k}} Y^{i}\,\bar{\theta}^{k} = D_{k}\, Y^{i}\,\alpha^{k}. + D_{\dot{k}}\, Y^{i}\,\theta^{k}$$

where $D_{k}Y^{i}$ and $D_{\dot{k}}\, Y^{i}$ are two tensors in the large sense of the type (1. I). On putting in this relation the $\bar{\theta}^{k}$ defined by (3.2), we obtain

(3.7) $$D_{k}Y^{i} = \bar{\nabla}_{k}\, Y^{i} + \bar{\nabla}_{\dot{h}}\, Y^{i}\, S^{h}_{ok}$$

(3.8) $$D_{\dot{k}} Y^i = \overline{\nabla}_{\dot{k}} Y^i$$

b. Let U be a neighborhood of V(M) consisting of ordinary points . Let us consider a local section invariant by X over $\overline{U}$. We have shown that for such a section we have (2.4), namely

(3.9) $$[L(X)\alpha]^i = L(X)\,\alpha^i = i\,(\hat{X})d\alpha^i + di(\hat{X})\alpha^i = 0$$

Taking into account [7.3 chap I], this relation becomes

(3.10)
$$dX^i + (\omega^i_j + S^i_{jk}\alpha^k)X^j - \omega^i_k(\hat{X})\alpha^k + T^i_{kh}\,\dot{X}^h\,\alpha^k - X^h\,T^i_{hk}\,\theta^k = 0$$
Let

(3.11) $$\overline{\nabla} X^i + T^i_{kh}\dot{X}^h\alpha^k = \omega^i_k(\hat{X})\alpha^k + X^h\,T^i_{hk}\,\theta^k$$

In virtue of (3.6), for the frames of a local section in question we deduce

(3.12) $$D_k X^i + T^i_{kh}\dot{X}^h = \omega^i_k(\hat{X})$$

(3.13) $$D_{\dot{k}} X^i = X^h\,T^i_{hk}$$

c. It is now easy to evaluate the Lie derivative of a tensor in the large sense or of a regular linear connection. First let us calculate the infinitesimal transformation of a vector field Y in the large sense. Given a neighborhood $\overline{U}$ consisting of ordinary points, let us consider a local section invariant of X over $\overline{U}$. With respect to the frames of this section, Y defines a 0-form with vector values $Y\overline{U} = (Y^i)$ of (2.3), we then have for these frames

$$[L(X)Y)]^i = L(X)\,Y^i = i(\hat{X})\,dY^i = \nabla Y^i(\widetilde{X}) - Y^k\,\omega^i_k(\hat{X})$$

Taking into account (3.12), it becomes

(3.14) $$[L(X)Y)]^i = \nabla Y^i(\hat{X}) - Y^k(D_k X^i + T^i_{kh}\dot{X}^h)$$

the two sides of this formula being vectors in the large sense on V(M) the equality is valid for arbitrary frames and at the ordinary points of V(M). Passing to the limit it is valid at all the points. In order to evaluate the quantities $\dot{X}$ that figure in the right hand side of (3.14), it suffices to remark the Lie derivative of the canonical vector field v is zero, whence

(3.15) $$\dot{X}^i = D_o X^i + T^i_{oh}\dot{X}^h$$

Let

(3.16) $$\dot{X}^h(\delta^i_h - T^i_{oh}) = D_0 X^i$$

On the other hand, the formula (3.14) can be put under the form

$$[L(X)Y)]^i = X^k \nabla_k Y^i - Y^k D_k X^i + \dot{X}^k(\nabla_{\dot{k}} Y^i - Y^h T^i_{hk})$$

where the $\dot{X}$ are defined by (3.16). In a local co-frame (dx, θ) the term within the parenthesis in the right hand side of the formula is none other than the Pfaffian derivative of Y^i with respect to θ^k; on taking into account [(6.9), chap. I] and (3.16), we obtain

(3.17) $$[L(X)Y)]^i = X^k \nabla_k Y^i - Y^k D_k X^i + \delta_{\dot{s}} Y^i D_0 X^s.$$

The right hand is independent of the connection introduced. Let us suppose that the coefficients of ω with respect to the co-frame (dx, dv) are zero. Then it will be so also for the associated connection, and the relation (3.17) is reduced to

(3.18) $$[L(X)\,Y)]^i = X^k \delta_k Y^i - Y^k \delta_k X^i + \delta_{\dot{s}} Y^i \delta_0 X^s.$$

4. The Lie Derivative of a Tensor in the large sense.

We will now apply the above method to calculate the Lie derivative of a tensor in the large sense. Let t^i_{jk} be a tensor in the large sense on V(M); in the frames of a local section, invariant by X, we have

$$[L(X)t]'^{i}_{jk} = i(\hat{X})dt'^{i}_{jk}$$
$$= \nabla t'^{i}_{jk}(\hat{X}) + t'^{i}_{rk}\omega'^{r}_{j}(\hat{X}) + t'^{i}_{jr}\omega'^{r}_{k}(\hat{X}) - t'^{r}_{jk}\omega'^{i}_{r}(\hat{X})$$

In virtue of (3.12) we obtain

$$(4.1)\qquad [L(X)t]'^{i}_{jk} = \nabla t'^{i}_{jk}(\hat{X}) + t'^{i}_{rk}(D_jX^r + T'^{r}_{jh}\dot{X}^h) + t'^{i}_{jr}(D_kX^r + T'^{r}_{kh}\dot{X}^h) - t'^{r}_{jk}(D_rX^i + T'^{i}_{rh}\dot{X}^h)$$

where $\dot{X}$ is defined by (3.16). This formula is valid in arbitrary frames and at all points. By the same reasoning as above, we have in local co-frame (dx^i, θ^i)

$$(4.2)\qquad [L(X)t)]'^{i}_{jk} = X^r\,\nabla_r\, t'^{i}_{jk} + t'^{i}_{rk}\, D_j\, X^r + t'^{i}_{jr}D_k\, X^r - t'^{r}_{jk}\, D_r\, X^i + \delta^{\bullet}_{s}\, t'^{i}_{jk}D_oX^s$$

and

$$(4.3)\qquad [L(X)t]'^{i}_{jk} = X^r\delta_r\, t'^{i}_{jk} + t'^{i}_{rk}\delta_jX^r + t'^{i}_{jr}\,\delta_k\, X^r - t'^{r}_{jk}\delta_r\, X^i + \delta^{\bullet}_{s}\, t'^{i}_{jk}\,\delta_oX^s$$

5. The Lie derivative of the coefficients of a regular linear connection.

Let ω be a regular linear connection on V(M). We know that the Lie derivative of ω is a tensorial 1-form on $p^{-1}E(M)$. of the type adj. Let $\overline{U}$ be a neighborhood consisting of ordinary points. Let us consider a local section invariant under X over $\overline{U}$. After (2.3), in terms of these local frames we have

$$(5.1)\qquad [L(X)\,\omega]'^{i}_{j} = L(X)\,\omega'^{i}_{j} = i(\hat{X})\, d\omega'^{i}_{j} + di(\hat{X})\omega'^{i}_{j}$$

The operator i is an anti-derivation from [(7.6), chap I]; so we obtain

(5.2) $i(\hat{X})\, d\omega^i_j = i(\hat{X})\Omega^i_j - \omega^i_r(\hat{X})\omega^r_j + \omega^i_r\,\omega^r_j(\hat{X})$

In virtue of [(7.7), chap I], the first term of the right hand side becomes

(5.3) $i(\hat{X})\Omega^i_j = (X^h R^i_{jhk} - \dot{X}^h P^i_{jkh})\alpha^k + (X^h P^i_{jhk} + \dot{X}^h Q^i_{jhk})\,\theta^k$

With respect to the frames in question we have, after (3.12),

(5.4) $di(\hat{X})\omega^i_j + \omega^i_r\,\omega^r_j(\hat{X}) - \omega^i_r(\hat{X})\omega^r_j = \nabla(D_j X^i + T^i_{jh}\dot{X}^h)$

Taking into account (5.2), (5.3) and (5.4), the formula (5.1) becomes

$$\begin{aligned}(5.5)\qquad [L(X)\,\omega]^i_j = L(X)\omega^i_j &= \nabla(D_j X^i + T^i_{jh}\,\dot{X}^h)\\ &+ (X^h R^i_{jhk} - \dot{X}^h P^i_{jkh})\alpha^k + (X^h P^i_{jhk} + \dot{X}^h Q^i_{jhk})\theta^k\end{aligned}$$

where $\dot{X}$ is defined by (3.16). Relative to the above frames, the Lie derivative of θ^r becomes

$$\begin{aligned}(5.6)\qquad [L(X)\,\theta]^r = L(X)\,\theta^r &= L(X)(dv^r + v^s\,\omega^r_s)\\ &= [L(X)\,\omega]^r_o\end{aligned}$$

Taking into account this relation, the formula (5.5) becomes in local co-frame (dx, θ),

$$\begin{aligned}(5.7)\qquad [L(X)\,\omega]^i_j = L(X)\overset{*}{\Gamma}{}^i_{jk}\,dx^k &+ L(X)T^i_{jk}\theta^k\\ &+ T^i_{jr}[L(X)\omega]^r_o\end{aligned}$$

whence, on multiplying the two sides by v^j

$$[L(X)\,\omega]^i_o = L(X)\overset{*}{\Gamma}{}^i_{ok}\,dx^k + L(X)\,T^i_{ok}\,\theta^k + T^i_{or}[L(X)\omega]^r_o$$

From it, we have

$$(5.8)\qquad [L(X)\,\omega]^r{}_o = M^r{}_s\,(L(X)\overset{*}{\Gamma}{}^s{}_{ok}dx^k + L(X)\,T^s{}_{ok}\theta^k)$$

where M is the inverse of the matrix L. On substituting (5.8) in (5.7) and taking into account (5.5) we get

$$(5.9)\quad [L(X)\omega]^i{}_j = (L(X)\overset{*}{\Gamma}{}^i{}_{jk} + T^i{}_{jr}M^r{}_sL(X)\overset{*}{\Gamma}{}^s{}_{ok})dx^k + (L(X)T^i{}_{jk} + T^i{}_{jr}M^r{}_sL(X)T^s{}_{ok})\theta^k$$

$$= \nabla(D_jX^i + T^i{}_{jh}\dot{X}^h) + (X^hR^i{}_{jhk} - \dot{X}^hP^i{}_{jkh})dx^k + (X^hP^i{}_{jhk} + \dot{X}^hQ^i{}_{jhk})\theta^k$$

In virtue of (5.6), the formula (5.5) gives us
(5.10)

$$\begin{aligned} L(X)\overset{*}{\Gamma}{}^i{}_{jk}dx^k + L(X)\,T^i{}_{jk}\theta^k &= \nabla(D_jX^i + T^i{}_{jh}\dot{X}^h) + (X^h R^i{}_{jhk} - \dot{X}^hP^i{}_{jkh})dx^k \\ &+ (X^hP^i{}_{jhk} + \dot{X}^hQ^i{}_{jhk})\theta^k - T^i{}_{jr}v^s\,[\nabla(D_sX^r + T^r{}_{sh}\dot{X}^h) \\ &+ (X^hR^r{}_{shk} - \dot{X}^hP^r{}_{skh})dx^k + (X^hP^r{}_{shk} + \dot{X}^hQ^r{}_{shk})\theta^k] \end{aligned}$$

where $\dot{X}$ is defined by (3.16). The two sides of the formulas (5.9) and (5.10) being tensor 1-forms of type adj, these relations are valid in arbitrary frames and at all points, whence

$$\begin{aligned} (5.11)\quad L(X)\overset{*}{\Gamma}{}^i{}_{jk} &= \nabla_k(D_jX^i + T^i{}_{jh}\dot{X}^h) + (X^hR^i{}_{jhk} - \dot{X}^hP^i{}_{jkh}) \\ &- T^i{}_{jr}v^s\,[\nabla_k(D_sX^r + T^r{}_{sh}\dot{X}^h) + (X^hR^r{}_{shk} - \dot{X}^hP^r{}_{skh}] \end{aligned}$$

and

$$\begin{aligned} (5.12)\quad L(X)\,T^i{}_{jk} &= \nabla_{\dot{k}}\,(D_jX^i + T^i{}_{jh}\dot{X}^h) + (X^hP^i{}_{jhk} + \dot{X}^h\,Q^i{}_{jhk}) \\ &- T^i{}_{jr}v^s[\nabla_{\dot{k}}\,(D_sX^r + T^r{}_{sh}\dot{X}^h) + (X^hP^r{}_{shk} + \dot{X}^hQ^r{}_{shk})] \end{aligned}$$

In virtue of (7.26), (7.27) of Chapter I, the formula (5.11) becomes

(5.13) $$L(X)\overset{*}{\Gamma}{}^{i}{}_{jk} = \nabla_k D_j X^i + X^h R^{*i}{}_{jhk} + \delta^{\bullet}_{s} \overset{*}{\Gamma}{}^{i}{}_{jk} D_o X^s$$

As to (5.12), on developing the right hand side we re-find the formula giving us the Lie derivative of the tensor T (4.2).

6. Fundamental Formula. –In the following we suppose that V(M) is endowed with a regular linear connection and we reason in local co-frames (dx^i, θ^i). Let g the vector space over **R** of infinitesimal transformations defined by the vector fields on M. To every pair $(X, Y) \in g$, we let correspond the usual bracket

(6.1) $$[X,Y] = L(X)Y\ ;$$

With this composition law g becomes a Lie algebra, which will be called the Lie algebra of infinitesimal transformations of M, and will be denoted by g. Let us consider the map which lets correspond to every $X \in g$ the tensor of the type adj defined by

(6.2) $$(AX)^i{}_j = -\,(D_j X^i + T^i{}_{jh}\,\dot{X}^h),$$

where

(6.3) $$\dot{X}^h\,(\,\delta^r{}_h - T^r{}_{oh}) = D_o X^r$$

In local co-frames (dx, θ) the expression between parentheses in (6.3) is none other than the matrix $L^r{}_h$ defined by [(5.8), chap.I], the linear connection in question being regular; this matrix has therefore an inverse which we have denoted by M; from (6.3) we obtain

(6.4) $$\dot{X}^h = M^h{}_r\,D_o X^r$$

On substituting (6.4) in (6.2), we have

$$(AX)^i{}_j = -(D_j X^i + T^i{}_{jh} M^h{}_r D_o X^r)$$

Thus to every $X \in g$ we let correspond the endomorphism Az(X) of $T_{\pi z}$ We denote by Az(g) the vector space of endomorphisms corresponding to all the elements of g. Az(g) can thus given the structure of a Lie algebra by the composition law defined by the usual bracket : if AX and AY are two elements of Az(g), their bracket is defined by

$$[AX, AY] = AXAY - AYAX. \tag{6.5}$$

We will denote this algebra by A. Let X be an element of g and AX the corresponding endomorphism: if t^i_{jk} is a tensor field in the large sense, we put

$$(AXt)^i_{jk} = (AX)^i{}_r\, t^r_{jk} - (AX)^r_j\, t^i_{rk} - (AX)^r_k\, t^i_{jr} \tag{6.6}$$

With these notations, the formula (4.1), giving the Lie derivative of t^i_{jk} by X can be written in a simple manner

$$[L(X)t]^i_{jk} = \nabla t^i_{jk}(\widetilde{X}) + (AXt)^i_{jk} \tag{6.7}$$

Let X and Y be two elements of g and AX and AY the corresponding elements of A; to the bracket [X, Y] corresponds the element A[X,Y] of A ; we propose to evaluate A[X, Y] as a function of AX and AY. If we put

$$Z = [X, Y],$$

we have by (6.2)

$$(AZ)^i_j = -(D_jZ^i + T^{\,i}_{\ jh}\, \dot{Z}^{\,h}) \tag{6.8}$$

with

$$\dot{Z}^{\,h} = D_o Z^h + T^h_{ok}\, \dot{Z}^{\,k} \tag{6.9}$$

From (6.7) it follows

$$[L(X)AY]^i_j = \nabla (AY)^i_j(\hat{X}) + [AX, AY]^i_j$$

Taking into account the definition of AY, this relation can be written

(6.10) $$[AX, AY]^i_j + \nabla(AY)^i_j(\hat{X}) = - [L(X)D_j Y^i + L(X)T^i_{jh}\dot{Y}^h + T^i_{jh}L(X)\dot{Y}^h]$$

where

(6.11) $$\dot{Y}^h = D_0 Y^h + T^h_{ok}\dot{Y}^k.$$

On subtracting (6.10) from (6.8), we get

(6.12)

$$(AZ)^i_j - [AX, AY]^i_j - \nabla(AY)^i_j(\widetilde{X}) = L(X)D_j Y^i - D_j L(X) Y^i + T^i_{jh}(L(X)\dot{Y}^h - \dot{Z}^h) + L(X) T^i_{jh} Y^h$$

We are going to evaluate the right hand side of this relation; first of all, in virtue of (3.7), we have

$$D_j Y^i = \overline{\nabla}_j Y^i + Y^m T^i_{mh} S^h_{0j} = \delta_j Y^i + Y^k \overset{*}{\Gamma}{}^i_{jk}$$

whence

(6.13) $$L(X) D_j Y^i - D_j L(X) Y^i = Y^k L(X) \overset{*}{\Gamma}{}^i_{jk}$$

On the other hand, on taking the Lie derivative with respect to X of the two sides of (6.11) we have

$$L(X)\dot{Y}^h = L(X)D_o Y^h + L(X) T^h_{ok}\dot{Y}^k + T^h_{ok} L(X)\dot{Y}^k$$

On substituting in this relation the expression of $\dot{Z}^h$ defined by (6.9) and on taking into account (6.13), we obtain

$$L(X)\dot{Y}^h - \dot{Z}^h = T^h_{ok}(L(X)\dot{Y}^k - \dot{Z}^k) + Y^k L(X)\overset{*}{\Gamma}{}^h_{ok} + \dot{Y}^k L(X)T^h_{ok}$$

whence

(6.14) $$L(X)\dot{Y}^h - \dot{Z}^h = M^h_r(Y^k L(X)\overset{*}{\Gamma}{}^r_{0k} + \dot{Y}^k L(X) T^r_{0k}).$$

In virtue of (6.13) and (6.14), the relation (6.12) can be written

$$(AZ)^i_j - [AX, AY]^i_j - \nabla(AY)^i_j(\hat{X})$$
$$= Y^k\,(L(X)\,\overset{*}{\Gamma}{}^i_{jk} + T^i_{jh}\,M^h_r\,L(X)\overset{*}{\Gamma}{}^r_{0k})$$
$$+\ \dot{Y}^k(L(X)\,T^i_{jk} + T^i_{jh}M^h_r\,L(X)T^r_{ok}).$$

On evaluating the right hand side of this relation by (5.9), we obtain the fundamental formula

$$(6.15)\quad A[X,Y] = [AX, AY] + \nabla AY(\hat{X}) - \nabla AX(\hat{Y}) + \Omega(\hat{X},\hat{Y}).$$

7. Divergence Formulas

When the manifold M is compact we have established in [1] a divergence formula for a horizontal 1-form. Using the same calculus, we establish here an analogous formula for a vertical 1–form (See [2]). Let u: M → W(M) a unitary vector field. We denote by $u = u_i\, dx^i$ the 1-form associated to u. Now du is of rank 2(n-1) everywhere on W(M). We denote by $(du)^{n-1}$ the $(n-1)^{th}$ exterior power of du, by η the volume element of W(M):

$$\eta = \frac{(-1)^N}{(n-1)!}\Phi, \qquad \Phi = u \wedge (du)^{n-1} \qquad N = \frac{n(n-1)}{2}$$

Let $(x, e_1, e_2, \ldots e_{n-1}, u)$ be an ortho-normal frame of $T_{\pi y}$ $(y \in W)$, $(u = e_n)$

Then we have

$$u = u_i\, e_i = e_n \Rightarrow u_\alpha = 0, \qquad u_n = 1$$

Let $\beta_i = \nabla u_i$ ((3.2) chap II); then we have

$$\beta_n = \omega_{nn} = 0, \quad \beta_\lambda = \omega_{\lambda n}, \qquad (\lambda = 1, \ldots n\text{-}1)$$

Relative to this frame the volume element η becomes

$$\eta = \beta \wedge \omega$$

where

$$\beta = \beta_1 \wedge \ldots \wedge \beta_{n-1}, \quad \omega = \omega_1 \wedge \ldots \wedge \omega_n$$

with (ω_i) is the dual co-frame of (e_i) Let b be a vertical 1-form on W

$$b = \sum_{\lambda=1}^{n-1} b_\lambda \beta_\lambda \qquad (b_n = 0) \tag{7.1}$$

The adjoint of b relative to η is

$$*b = \sum_{\lambda=1}^{n-1} (-1)^{\lambda-1} b_i \beta_i \wedge \hat{\beta}_\lambda \wedge \quad \wedge \beta_{n-1} \wedge \omega \tag{7.2}$$

where the sign $\hat{\beta}$ signifies that the corresponding term should be omitted in the exterior product under consideration. Hence we have

$$\begin{aligned} (7.3) \quad d(*b) = & \sum_{\lambda=1}^{n-1} (-1)^{\lambda-1} db_\lambda \wedge \beta_1 \wedge \ldots \wedge \hat{\beta}_\lambda \wedge \quad \wedge \beta_{n-1} \wedge \omega \\ & \sum_{\lambda=1}^{n-1} (-1)^{\lambda-1} b_\lambda \sum_{\mu \neq \lambda} (-1)^{\mu-1} d\beta_\mu \wedge \beta_1 \wedge \quad \wedge \hat{\beta}_\mu \wedge \wedge \hat{\beta}_\lambda \wedge \beta_{n-1} \wedge \omega \\ & + (-1)^{n-2} \sum_{\lambda=1}^{n-1} (-1)^{\lambda-1} b_\lambda \beta_1 \wedge \hat{\beta}_\lambda \wedge \quad \wedge \beta_{n-1} \wedge d\omega \end{aligned}$$

Therefore we should calculate the terms $d\beta_\mu$ and $d\omega$. First of all we have

$$d\beta_i = \beta_h \wedge \omega_{ih} + \frac{1}{2} F^{-1} R_{iokl} \omega_k \wedge \omega_l + P_{iokl}\, \omega_k \wedge \beta_l$$

whence, modulo the terms containing the ω_k

$$d\beta_i \equiv \beta_h \wedge \omega_{ih} \qquad (\text{mod } \omega_k) \tag{7.4}$$

Similarly

$$d\omega = \sum_{i=1}^{n} (-1)^{i-1} d\omega_i \wedge \omega_1 \wedge \ldots \wedge \hat{\omega}_i \wedge \ldots \wedge \omega_n$$

Now

$$d\omega_i = \omega_{ji} \wedge \omega_j = F\, T_{ijh}\, \beta_h \wedge \omega_j$$

Thus

(7.5) $$d\omega = F\, T_{iih}\, \beta_h \wedge \omega$$

On substituting (7.4) and (7.5) in (7.3) we obtain

$$d(*b) = F\,[\nabla_{\dot{\lambda}}\, b_\lambda + b_\lambda T_{ii\lambda}]$$

Where $\nabla_{\dot{\lambda}}$ denotes the covariant derivative with respect to the co-frame ∇v
Now,

$$u^i \nabla_{\dot{i}}\, b_n = 0 \Rightarrow u^n\ \nabla_{\dot{n}}\ b_n = 0 = \nabla_{\dot{n}}\ b_n$$

So

(7.6) $$\delta b = -F\,(\nabla_{\dot{i}}\, b_i + b_j\, T_{iij})$$

where δ is the co-differential (see [1], [2]) Let us suppose M to be compact without boundary. We then have

(7.7) $$\int_{W(M)} \delta b \eta = - \int_{W(M)} F(\nabla_{\dot{i}}\ b^i + b^i\, T_i)\eta = 0$$

To this relation we give the name the divergence formula for a vertical 1-form (7.1). Similarly, if

(7.8) $$a = a_i(z)\, dx^i, \qquad z \in V(M)$$

is a horizontal 1-form on W(M) we obtain (see [1] pp 68-70)

(7.9) $$\delta a = -\,(\nabla_i a^i - a_i \nabla_0 T^i)$$

In case M is compact without boundary, we have

(7.10) $$\int_{W(M)} \delta\, a\, \eta = - \int_{W(M)} (\nabla_i a^i - a_i \nabla_0 T^i)\eta = 0$$

This is the divergence formula for a horizontal 1–form.

8. Infinitesimal Isometries [1a], [1b]

A. Let X be an infinitesimal transformation on M and H_z the horizontal sub-space of $T_Z V(M)$ defined by the Finslerian connection. The extended group $\exp(u\hat{X})$ maps H_z to a subspace of $T_{z(u)}(V(M))$. An infinitesimal transformation X leaves H_z invariant if the image of H_z by $\exp(u\hat{X})$ is a horizontal subspace of $T_{z(u)}$.

Theorem 1. *In order that the infinitesimal transformation X leaves the horizontal distribution H_z invariant it is necessary and sufficient that it satisfies one of the three following equivalent conditions*:

(1) L(X) commutes with the projection H

(2) The splitting of the 1-form ∇v is invariant by X

(3) *The bracket $[\hat{X}, \hat{v}]$ is zero, where $\hat{v}$ is the horizontal vector field above v ($p_* \hat{v} = v$)*

(1) is evident. To show that (1) implies (2) we calculate the Lie derivative of ∇v. On identifying ω in (3.1) with the Finslerian connection (S = 0) and on multiplying the two sides of (5.11) by v^j we obtain

(8.1) $$L(X)\, \nabla v(H\hat{Y}) = \nabla_{H\hat{Y}} \nabla_{\hat{v}}(X) + R(X, Y)v + (\nabla_{\hat{v}} T)(\nabla_{\hat{v}} X, Y)$$

Now $[\hat{X}, H\hat{Y}]$ is, by hypothesis, horizontal. This implies that the right hand side vanishes.

(2) $\Rightarrow$ (3): it suffices to take $\hat{Y} = \hat{v}$ in (8.1) and to remark that $[\hat{X}, \hat{v}]$ is always vertical from (8.1) the theorem follows.

(8.2) $$\nabla_{\hat{v}} \nabla_{\hat{v}} X + R(X, v)\, v = 0 \Leftrightarrow [\hat{X}, \hat{v}] = 0$$

(3)$\Rightarrow$ (2) In fact we have noted for $2\,G^i$ the expression $\overset{*}{\Gamma}{}^i{}_{jk} v^j v^k$ (see 7.12 chapter II). Now the left hand side of (8.2) is none other than the Lie derivative of $2G^i$ (see 5.11) So, on deriving vertically we obtain (2).

B. Let X be an infinitesimal transformation on M. Then X is an infinitesimal isometry if it leaves invariant the metric tensor g:

(8.3) $$L(X)g = 0$$

Let us decompose L(X):

$$L(X) = \nabla_{\hat{X}} + A_{\hat{X}}$$

where $A_{\hat{X}}$ is defined by

(8.4) $$A_{\hat{X}}\, Y = -[\nabla_{H\hat{Y}} X + T(\nabla_{\hat{v}} X, Y)]$$

Theorem 2. *X is an infinitesimal isometry if it satisfies one of following three equivalent conditions:*

(1) $$g(A_{\hat{X}}\, Y, Z) + g(Y, A_{\hat{X}}\, Z) = 0$$

(2) $$g(\nabla_{H\hat{Y}} X, v) + g(Y, \nabla_{\hat{v}} X) = 0$$

(3) $$g(\nabla_{\hat{v}} X, v) = 0$$

for all $Y, Z \in p^{-1}T(M)$ where g(,) denotes the local scalar product.

(1) signifies that the Lie derivative of g is zero. (1) $\Rightarrow$ (2) $\Rightarrow$ (3) is evident. To prove the opposite implication we must derive vertically

The compact case

Theorem 3. *For a compact manifold without boundary M the largest connected group of transformations which leave invariant the splitting defined by the Finslerian connection coincides with the largest connected group of isometries([1a], [1b])*
Proof. If X is an infinitesimal isometry, it leaves invariant the 1-form of the splitting ∇. Hence (8.2). Conversely for X satisfying (8.2) we obtain

$$(8.5) \qquad \nabla_{\hat{u}}\ [g(X, u)\, g(A_{\hat{X}}\, u\, u)] + g(A_{\hat{X}}\, u, u)^{2} = 0$$

where $\hat{v} = L\ \hat{u}$ $(L = \|v\|)$ and $v = Lu$. M being compact without boundary and on integrating (8.5) on W(M) the first term is a divergence. So, after the formula(7.10), we obtain

$$\int_{W(M)} g(A_{\hat{X}}\, u, u)^2\, \eta = 0$$

whence $g(A_{\hat{X}}\, u, u) = 0$
This is the condition 3 of theorem 2.

C. Let X be an infinitesimal isometry and $f = \frac{1}{2} g(X, X)$. On using (8.2) we obtain

$$\hat{u}\ \hat{u}\ (f) = g\ (\nabla_{\hat{u}} X,\ \nabla_{\hat{u}} X) - g(R(X, u)u, X)$$

Since M is compact and without boundary, the left hand side is a divergence. So integrating over W(M) :

$$(8.6) \qquad <(\nabla_{\hat{u}} X,\ \nabla_{\hat{u}} X> = \int_{W(M)} (R(X, u)u, X)\, \eta$$

where < , > and (,) denote respectively the global and local scalar product on W(M).

Let us suppose that the integral of the right hand side of (8.6) is non-positive. We then have $\nabla_{\hat{u}} X = 0$. From it we obtain, by vertical derivation, and the fact that X is an isometry, X is of covariant derivation of horizontal type zero. If the quadratic form (R(X, u)u, X) is negative definite the group reduces to the identity.

Theorem 4. *Let (M, g) be a compact Finslerian manifold without boundary and X an infinitesimal isometry. If the integral of (R(X,u)u,X) on W(M) is non-positive, X is of covariant derivation of horizontal type zero. If this form is negative definite then the group of isometry of (M, g) is a finite group ([1a], [1b]).*

9. Ricci Curvatures and Infinitesimal Isometries

In the preceding section we have shown the influence of the sign of the sectional curvature (R(X, u)u, X) (the flag curvature) on the existence of a non-trivial isometry group. Here we will highlight the significance of the Ricci curvatures R_{ij} and P_{ij} of the Finslerian connection. After theorem 2 of the previous paragraph, to every infinitesimal isometry X is associated an anti-symmetric endomorphism AX of T_{pz} defined by

(9.1) $$(A_X)^i{}_j = -(\nabla_j X^i + T^i{}_{jh} \nabla_o X^h)$$

To this endomorphism is associated a 2-form (A_X)

(9.2) $$A_X = \frac{1}{2}(\nabla_i X_j - \nabla_j X_i)\, dx^i \wedge dx^j$$

X being an isometry, the Finslerian connection is invariant under X by (5.11). So we have

(9.3) $$\nabla_k (A_X)^i{}_j = X^h R^i{}_{jhk} - P^i{}_{jkh} \nabla_o X^h$$

whence

(9.4) $$-\nabla^j (\nabla_j X_i - \nabla_i X_j) = 2\,(R_{ij} X^j + P_{ij} \nabla_o X^j)$$

where we have put $R^r{}_{irj} = R_{ij}$, and $P^r{}_{irj} = P_{ij}$ Then by (8.9 chap II)

$$(9.5) \qquad P_{ij} = \nabla_r T^r{}_{ij} - \nabla_i T_j + T_r \nabla_o T^r{}_{ij} - T^r{}_{si} \nabla_o T^s{}_{rj}$$

from (9.4) and using the divergence formula (7.10) we obtain

$$(9.6) \quad \frac{1}{2}\delta(i(X)A_X) = \frac{1}{2}(A_X, A_X) - R_{ij} X^i X^j - P_{ij} X^i \nabla_o X^j$$
$$+ \frac{1}{2} X^i(\nabla_i X_j - \nabla_j X_i)\nabla_o T^j$$

where i(X) is the inner product with X. We are going to calculate the last two terms of the right hand side when X is an isometry. Let us put the result in a quadratic form in X modulo divergences. After (9.5) we have, on taking into account the divergence formulas (7.7) and (7.9) :

the term: $-T^r{}_{si} X^i \nabla_o T^s{}_{rj} \nabla_o X^j = \text{Div}$
$$+[\nabla_o(T^r{}_{si} \nabla_o T^s{}_{rj}) + \frac{1}{2}(\nabla_r \nabla_o T^r{}_{ij} - \nabla_o T^r{}_{ij} \nabla_o T_r)]X^i X^j$$

the term $\nabla_r T^r{}_{ij} X^i \nabla_o X^j = \text{Div} - \frac{1}{2}(\nabla_o \nabla_r T^r{}_{ij}) X^i X^j$

the term $T_r \nabla_o T^r{}_{ij} X^i \nabla_o X^j = \text{Div} - \frac{1}{2} \nabla_o(T_r \nabla_o T^r{}_{ij}) X^i X^j$

the term : $-\nabla_i T_j X^i \nabla_o X^j$

$$(9.7) \qquad = \nabla_o \nabla_i T_j X^i X^j + \nabla_i T_j \nabla_o X^i X^j + \text{div on } W(M)$$

where div = divergence as in the rest of the book. Take the term in the expression: $\nabla_i T_j = D_i T_j + T_r \nabla_o T^r{}_{ij}$ where ∇ is the covariant derivation in the Finslerian connection and D is the covariant derivation in the Berwald connection. So we get

$$(9.8) \quad \nabla_i T_j \nabla_o X^i X^j = D_i T_j \nabla_o X^i X^j + \frac{1}{2} T_r \nabla_o T^r{}_{ij} \nabla_o (X^i X^j)$$
$$= \text{Div on } W(M) - \frac{1}{2} \nabla_o(T_r \nabla_o T^r{}_{ij}) X^i X^j + D_i T_j \nabla_o X^i X^j$$

But the last term of the right hand side is

$$D_i T_j \nabla_o X^i X^j = g^{ik} D_o X_k \; \dot{\partial}_i \; D_o T_j X^j - g^{ik} D_o X_k D_o \; \dot{\partial}_i \; T_j X^j$$

$$= g^{ik} \; \dot{\delta}_i \; [D_o X_k D_o T_j X^j]$$

$$- g^{ik} (D_i X_k + D_o \dot{\partial}_i X_k) D_o T_j X^j - D_o \dot{\partial}_i \; T_j X^j D_o X^i$$

Now $D_o X_o = 0$ since X is an isometry. The first term of the right hand side is a divergence. So we get

$$D_i T_j \nabla_o X^i X^j = \text{div on } W(M) + g^{ik} D_i X_k D_o T_j X^j$$

$$+ \frac{1}{2} D_o (D_o \dot{\partial}_i \; T_j) X^i X^j$$

$$= -D_o(X, T) D_o T_j X^j + \frac{1}{2} [D_o \dot{\partial}_i \; D_o T_j - D_o \nabla_i T_j$$

$$+ \nabla_o (T_r \nabla_o T^r{}_{ij})] X^i X^j + \text{div on } W(M)$$

On taking into account this relation, (9.8) becomes

$$\text{(9.9)} \quad \nabla_i T_j \nabla_o X^i X^j = \frac{1}{2} (D_o \dot{\partial}_i \; D_o T_j - D_o \nabla_i T_j) X^i X^j$$

$$D_o (X, T) D_o T_j X^j + \text{Div on } W(M)$$

We calculate the last term of the right hand side in another manner: X being an isometry we have

$$-\nabla_i \nabla_o T_j X^i X^j = -\nabla_i (\nabla_o T_j X^i X^j) + \nabla_i T_j \nabla_i X^i X^j + \nabla_o T_j X^i \nabla_i X^j$$

$$= X^j \nabla_o T_j \nabla_i X^i - X^j \nabla_o T_j X^i \nabla_o T_i$$

$$- \nabla_o T^j X^i (\nabla_j X_i + 2 T^h{}_{ij} \nabla_o X_h) + \text{Div on } W(M)$$

$$= - T^i \nabla_o X_i X^j \nabla_o T_j - X^j \nabla_o T_j X^i \nabla_o T_i$$

$$-\frac{1}{2} \nabla_o T^j \nabla_j X^i X_i - \nabla_o T_r T^r{}_{ij} \nabla_o (X^i X^j) + \text{Div on } W(M)$$

$$= - X^j \nabla_o T_j \nabla_o (X, T) - \frac{1}{2} \nabla_j (\nabla_o T^j X^i X_i) + \frac{1}{2} \nabla_j \nabla_o T^j X^i X_i$$

$$+ \nabla_o (\nabla_o T_r T^r{}_{ij}) X^i X^j + \text{Div on } W(M)$$

$$= \text{Div on } W(M) - X^j\nabla_o T_j \nabla_o (X, T) - \frac{1}{2} \nabla_o T^j \nabla_o T_j X^i X_i$$

$$+ \frac{1}{2} \nabla_j \nabla_o T^j X^i X_i + \nabla_o (\nabla_o T_r T^r_{ij})X^i X^j$$

$$= \text{Div on } W(M) + [\nabla_o (\nabla_o T_r T^r_{ij})$$

$$+ \frac{1}{2} (\nabla_r \nabla_o T^r - \nabla_o T^r \nabla_o T_r) g_{ij}]X^i X^j - X^j \nabla_o T_j \nabla_o (X,T)$$

Let us carry the expression - $X^j \nabla_o T_j \nabla_o (X, T)$, drawn from the preceding relation into (9.9):

$$-\nabla_i T_j \nabla_o X^i X^j = \frac{1}{2}(D_o\, \partial_i^\bullet D_o T_j - D_o \nabla_i T_j) X^i X^j - \nabla_i \nabla_o T_j X^i X^j$$

$$-[\nabla_o (\nabla_o T_r T^r_{ij}) + \frac{1}{2}(\nabla_r \nabla_o T^r - \nabla_o T^r \nabla_o T_r) g_{ij}]X^i X^j + \text{Div on } W(M)$$

This is the last term of the right hand side of (9.7). On substituting it in (9.7) we get:

$$-\nabla_i T_j X^i \nabla_o X^j = \text{Div} +$$

$$\{[\frac{1}{2} \nabla_o \nabla_i T_j - \nabla_i \nabla_o T_j + \frac{1}{2} \nabla_o \partial_i^\bullet \nabla_o T_j - \nabla_o (T_{ijh} \nabla_o T^h)$$

$$+ \frac{1}{2} g_{ij} [\nabla_o T^r \nabla_o T_r - \nabla_r \nabla_o T^r]\} X^i X^j$$

$$\text{the term } \frac{1}{2} X^i (\nabla_i X_j - \nabla_j X_i) \nabla_o T^j = \text{Div} + \frac{1}{2} [g_{ij} (\nabla_o T^r \nabla_o T_r - \nabla_r \nabla_o T^r)$$

$$+ \nabla_o (T_{ijh} \nabla_o T^h)]\, X^i X^j$$

Let us put

$$\psi(X, X) = \{\frac{1}{2} [(\nabla_o \nabla_r T^r_{ij} + \nabla_o \nabla_o (T_r T^r_{ij}) + \nabla_r \nabla_o T^r_{ij} - \nabla_r T^r_{ij} \nabla_o T_r -$$

$$\nabla_o \partial_i^\bullet \nabla_o T_j - \nabla_o \nabla_i T_j] - \nabla_o (T^r_{si} \nabla_o T^s_{rj}) + \nabla_o (T_{ijh} \nabla_o T^h) + \nabla_i \nabla_o T_j \} X^i X^j$$

We have

$$(9.10) \qquad - P_{ij}\, X^i \nabla_o X^j + \frac{1}{2} X^i (\nabla_i X_j - \nabla_j X_i) \nabla_o T^j$$
$$= \text{Div on } W(M) + \psi(X, X)$$

On substituting this relation in (9.6) we obtain

$$(9.11) \qquad \frac{1}{2}\delta(i(X)A_X) = \frac{1}{2}(A_X, A_X) - \varphi(X, X)$$

whence

$$(9.12) \qquad \varphi(X, X) = R_{ij} X^i X^j - \psi(X, X)$$

M being compact without boundary, on integrating on W(M) we get

$$(9.13) \qquad \frac{1}{2} < (A_X, A_X) > = \int_{W(m)} \varphi(X, X)\, \eta$$

By a reasoning analogous to the Riemannian case we show that the isometry group of a compact Finslerian manifold is compact since it is the isometry group of the manifold W(M) with the Riemannian metric of the fibre bundle associated to the Finslerian metric. Let us suppose that φ (X, X) in (9.13) is negative definite. It then follows that X vanishes so that the dimension of the isometry group is zero. Therefore it is a finite group. If $\varphi(X, X)$ in (9.13) is negative semi-definite it follows that X is of covariant derivation of horizontal type zero.

Theorem 5. *If the quadratic form $\varphi(X, X)$ is negative definite on W(M) then the isometry group of the compact Finslerian manifold without boundary is finite. If the form is negative semi-definite then X is of zero horizontal type covariant derivation [1b].*

If the Ricci tensor P_{ij} vanishes everywhere then by (9.10) $\psi(X, X)$ is a divergence. Hence the formula (9.13) reduces to

$$\frac{1}{2}(A_X, A_X) = \int_{W(m)} R_{ij}X^iX^j\eta$$

Corollary. *Let (M, g) be a compact Finslerian manifold without boundary such that the second Ricci tensor* P_{ij} *vanishes everywhere on W(M). If* $R_{ij}X^iX^j$ *is negative definite everywhere on W(M) then the isometry group of this manifold is finite. If this form is negative semi-definite then X is of zero horizontal covariant derivation [1b]..*

10. Infinitesimal Affine Transformations[1]

Let exp(uX) be the local 1-parameter group of local transformations of M, generated by X and its extended group $\exp(u\widetilde{X})$. Let ω be a regular linear connection V(M). We call infinitesimal affine transformation an infinitesimal transformation X on M such that its extended group $\exp(u\widetilde{X})$ leaves the linear connection ω invariant. For it to be so it is necessary and sufficient that

(10.1) $$L(X)\,\omega = 0$$

In virtue of (5.7) the above relation gives

(10.2) $$L(X)\,\dot{\Gamma}^{i}_{jk} = 0$$

(10.3) $$L(X)\,T^{i}_{jk} = 0$$

Conversely, if the relations (10.2) and (10.3) are satisfied for all X, then after (5.8) and (5.9), we have (10.1) and X defines an infinitesimal affine transformation.

For an infinitesimal transformation to define an affine infinitesimal transformation it is necessary and sufficient that it leaves invariant the coefficients of the connection. After (5.7) the relations (10.2) and (10.3) imply

(10.4) $\nabla(AX)^i_j = i(\widetilde{X})\Omega^i_j$

This is equivalent to

(10.5) $\nabla k\,(AX)^i_j = X^h R^i_{jhk} - \dot{X}^h P^i_{jkh}$

(10.6) $\nabla_{\dot{k}}\,(AX)^i_j = X^h P^i_{jhk} + \dot{X}^h Q^i_{jhk}$

where AX and $\dot{X}$ are defined respectively by (6.2) and (6.4). Conversely let us suppose that (10.4) is satisfied for all X, we then have, after (5.11) and (5.12), the relations (10.2) and (10.3) so that X is an affine infinitesimal transformation. We are thus, in the following, led to envisage the infinitesimal transformations that leave invariant the coefficients of the first kind $\dot{\Gamma}^i_{jk}$ of the connection. Such infinitesimal transformation will be called a *partial affine* infinitesimal transformation. In virtue of (5.9) and (5.10) , a partial infinitesimal transformation can be characterized by the system (10.5). Thus we have the

Theorem. *In order that an infinitesimal transformation X defines an affine infinitesimal transformation (respectively partial) for the connection ω, it is necessary and sufficient that it satisfies (10.4)[respectively (10.5)*)].[1]

It is clear that a partial affine infinitesimal transformation leaves invariant the geodesics of the linear connection ω. If X defines an affine infinitesimal transformation we have

(10.7) $L(X)\, R^i_{jkl} = 0, \qquad L(X)P^i_{jkl} = 0, \; L(X)Q^i_{jkl} = 0.$

11. Affine Infinitesimal Transformations and Covariant Derivations

Let t be a tensor of type R(G) with values in a vector space N and ω a regular linear connection on V(M). We will denote by $\widetilde{R}$(G) the representation of the Lie algebra <u>G</u> of G induced by $\widetilde{R}$.

Let $\overline{U}$ be a neighbourhood of V(M) consisting of ordinary points and a local section invariant under X over $\overline{U}$. The tensor t and the linear connection define on $\overline{U}$ respectively a 0-form $t_{\overline{U}}$ with values in N, a 1-form $\omega_{\overline{U}}$ with values in <u>G.</u> The absolute differential of t is a 1-form $\nabla t_{\overline{U}}$ with values in N

$$\nabla t_{\overline{U}} = \mathrm{d} t_{\overline{U}} + \widetilde{R}(\omega_{\overline{U}})\, t_{\overline{U}}$$

whence

$$L(X)\, \nabla t_{\overline{U}} = \mathrm{d}L(X)\, t_{\overline{U}} + \widetilde{R}(L(X)\, \omega_{\overline{U}})\, t_{\overline{U}} + \widetilde{R}(\omega_{\overline{U}}) L(X)\, t_{\overline{U}}$$

From this one deduces

$$L(X)\, \nabla t_{\overline{U}} - \nabla L(X)\, t_{\overline{U}} = \widetilde{R}(L(X)\omega)\, t_{\overline{U}}$$

Hence we have at every point of V(M)

$$L(X)\, \nabla t - \nabla L(X) t = \widetilde{R}(L(X)\omega) t \tag{11.1}$$

In particular, for a vector field in the large sense Y, we have

$$L(X)\, \nabla_k Y^i - \nabla_k L(X) Y^i = L(X)\, \dot{\Gamma}^i{}_{jk}\, Y^j - L(X)\, \dot{\Gamma}^r{}_{ok}\, \delta_{\dot{r}}\, Y^i$$
$$L(X)\, \nabla_{\dot{k}} Y^i - \nabla_{\dot{k}} L(X)\, Y^i = L(X)\, T^i{}_{jk}\, Y^j - L(X)\, T^r{}_{ok}\, \delta_{\dot{r}}\, Y^i$$

From the formulas (11.2) and (11.3) we have the following

Theorem. *In order that an infinitesimal transformation X defines an affine infinitesimal transformation (respectively partial) for the connection* ω*, it is necessary and sufficient that the covariant derivatives of the two types* ∇_k *and* $\nabla_{\dot{k}}$ *(respectively of type* ∇_k*) of a tensor in the large sense commutes with its Lie derivative by X [1].*

12. The Group Kz(L)

Let L be the Lie algebra of affine infinitesimal transformations of V(M) for the connection ω. In the preceding paragraphs we have seen how to every vector field $X \in L$ on M, we can let correspond a vector field $\widetilde{X}(X, \dot{X})$ on V(M) which is a lift of X. We denote by $\widetilde{L}$ the lift of L over V(M). On $\widetilde{L}$ we can define a Lie algebra structure by the usual bracket operation. The bracket of two elements of $\widetilde{L}$ can be projected by the map p. Let $\underline{A}z(L)$ be the Lie algebra of endomorphisms of $T_{\pi z}$ corresponding to the elements of L. $\underline{A}z$ (L) is the Lie algebra of a connected group Kz(L) of linear transformations of Tpz. We know that at a point $z \in V(M)$ the Lie algebra of infinitesimal holonomy can be generated by the elements(see [1], §11 chapter I)

(12.1) $$r'^i_j = \nabla\ldots\nabla R'^i_{jkl}(\widetilde{Z}_1 \ldots \widetilde{Z}_q) u'^k_1\, u'^l_2$$

(12.2) $$p'^i_j = \nabla\ldots \nabla P'^i_{jkl}(\widetilde{Z}_1 \ldots \widetilde{Z}_q) u'^k_1\, \dot{u}^l_2$$

(12.3) $$q'^i_j = \nabla\ldots\nabla Q'^i_{jkl}(\widetilde{Z}_1 \ldots \widetilde{Z}_q)\, \dot{u}^k_1\, \dot{u}^l_2$$

where $\widetilde{Z}_1 \ldots \widetilde{Z}_q$, $\widetilde{u}_1(u_1, \dot{u}_1)$ and $\widetilde{u}_2(u_2, \dot{u}_2)$ are the local vector fields in the neighbourhood of $z \in V(M)$. We denote by $\underline{\sigma}'z$ the Lie algebra of infinitesimal holonomy at $z \in V(M)$, by $\underline{r}'z$, $\underline{p}'z$, $\underline{q}'z$ those generated by r, p and q respectively. If $X \in L$, the Lie derivative with respect to X commutes with the covariant derivative and X leaves invariant the three curvature tensors, thus L(X)r, L(X)p, and L(X)q belong respectively to $\underline{r}'z$, $\underline{p}'z$, $\underline{q}'z$. On the other hand, in virtue of (6.7), the Lie derivative of r'^i_j by X (for example) becomes

$$L(X)\, r'^i_j = \nabla r'^i_j(\widetilde{X}) + [A_z(X), r]'^i_j$$

The left hand side as well as the first term of the right hand side belong to $\underline{r}'z$, whence

$$[A_z(X), \underline{r}'z\,] \subset \underline{r}'z$$

Similarly $$[A_z(X), \underline{p}'z\,] \subset \underline{p}'z \text{ and } [A_z(X), \underline{q}'z\,] \subset q'z$$

$\underline{\sigma}'z$ being the sum of $\underline{r}'z$, $\underline{p}'z$, $\underline{q}'z$, $[A_z(X), \underline{\sigma}'z] \subset \underline{\sigma}'z$

Thus we have the
Theorem. *The group Kz(L) is the subgroup of the connected normalizer $N_0(\sigma'z)$ of the group of infinitesimal holonomy at $z \in V(M)$ in the group of linear transformations of Tpz[1].*

13. Transitive algebra of affine infinitesimal transformations

1°. Let θz be the tangent space to V(M) at z

Definition. *The Lie algebra $\tilde{L}$ of affine infinitesimal transformations is said to be transitive on V(M) if at every point $z \in V(M)$ the subspace of θz generated by the $\tilde{X}z$ for $\tilde{X} \in \tilde{L}$ coincides with θz.*

In this paragraph we will suppose that $\tilde{L}$ is transitive on V(M). Let $z \in V(M)$ and let $\overline{U}$ and $\overline{V}$ be two neighborhoods of z. From the definition above it follows that for every point $\bar{z} \in \overline{U}$ there exists an affine transformation of $\overline{V}$ on a neighborhood of $\bar{z}$ that maps z on $\bar{z}$. From this we deduce that over $\overline{U}$ the groups of infinitesimal holonomies have the same dimension, hence z is regular[27] for the infinitesimal holonomy. Thus all the points of V(M) being regular for the infinitesimal holonomy, the group of infinitesimal holonomy $\underline{\sigma}'z$ at every point of $z \in V(M)$ coincides the restricted holonomy group σz

2°. Let X and Y be two elements of L, from (10.4) we have

$$\nabla AX(\tilde{Y}) = \Omega(\tilde{X}, \tilde{Y})$$

From (6.15) it follows

$$\Omega(\tilde{X}, \tilde{Y}) = [AX, AY] - A[X, Y]$$

whence, for every pair $X, Y \in L$, we have

$$\Omega(\widetilde{X}, \widetilde{Y}) \in \underline{A}z\,(L)$$

Let $Z_1 \in L$, from (6.7) we have

(13.1) $L(Z_1)\,\Omega(\widetilde{X}, \widetilde{Y}) = [A\,Z_1, \Omega)] + \nabla\Omega(\widetilde{X}, \widetilde{Y})(\widetilde{Z}_1)$

Let us put $[Z_1, X] = \xi_1,$ $[Z_1, Y] = \eta_1$

Since Z_1 belongs to L, we have

$$L(Z_1)\ \dot{X}^h = D_o\, L(Z_1)\, X^h + T^h{}_{ok}\, L(Z_1)\ \dot{X}^k$$

whence

$$L(Z_1)\,\dot{X}^h = M^h{}_k\, D_o\, \xi^h{}_1$$

The right hand side is none other than $\dot{\xi}_1^h$, whence

$$L(Z_1)\,\dot{X}^h = \dot{\xi}_1^h$$

Similarly $L(Z_1)\ \dot{Y}^h = \dot{\eta}_1^h$

Thus the left hand side of (13.3) can be written

$$L(Z_1)\,\Omega(\widetilde{X}, \widetilde{Y}) = \Omega(\widetilde{\xi}_1, \widetilde{Y}) + \Omega(\widetilde{X}, \widetilde{\eta}_1)$$

Whence, the left hand side is an element of $\underline{A}z(L)$, from the relation (13.3) it follows that $\nabla\Omega(\widetilde{X}, \widetilde{Y})(\widetilde{Z}_1)$ belongs to $\underline{A}z(L)$, on repeating this procedure we see therefore that the vector space Cq (q= 0, 1, 2, ...), generated by elements corresponding to the qth covariant derivation (q= m+n, m-times horizontal derivation and n-times vertical derivation) of the curvature tensors at z $\in V(M)$ belongs to $\underline{A}z(L)$ (See [1] page 20). Thus the restricted holonomy group σz is contained in the connected group Kz(L). Taking into the preceding theorem we have the

Theorem. *If the Lie algebra $\tilde{L}$ of affine infinitesimal transformations is transitive on V(M) at every point $z \in V(M)$ we have*

$$\sigma z \subset Kz(L) \subset N_0(\sigma z)$$

where $N_0(\sigma z)$ is the connected normaliser of the restricted holonomy group σz in the group of linear transformations of Tpz [1].

In the following L will be called basic Lie algebra of affine infinitesimal transformations.

14. The Lie Algebra L

1°. Let L be the basic Lie algebra of affine infinitesimal transformations and Az(L) the Lie algebra of endomorphisms of Tpz corresponding to the elements of L. The following lemma follows from the system of linear differential equations (10.4)

Lemma. *Let X and Y be two elements of L and AX and AY be the corresponding endomorphisms of Tpz, if at a point $x_0 = p\, z_0$ we have*

$$Xx_0 = Y x_0\,, \qquad Az_0(X) = Az_0\,(Y)$$

Then X and Y coincide on the entire M

After this lemma, the algebra L is finite dimensional.

2°. Let $\tilde{G}$ be the vector space defined by the direct sum

(14.1) $\tilde{G} = \underline{A}z(L) + Tpz$

On $\tilde{G}$ we define the following composition law :

(14.2) $[A_1, A_2] = A_1A_2 - A_2A_1$

$$[A, s] = -\,[s, A] = As$$

$$[s_1, s_2] = - \Sigma(\tilde{s}_1, \tilde{s}_2) - \Omega(\tilde{s}_1, \tilde{s}_2)$$

where A_1, $A_2 \in \underline{A}z(L)$, $s \in Tpz$, $A \in \underline{A}z(L)$, and $s_1, s_1 \in Tpz$ and $\tilde{s}$ is as usual defined by the pair $\tilde{s} = (s, \dot{s} = MD_0\, s)$

The composition law (14.2) does not in general define on $\tilde{G}$ the structure of a Lie algebra since the Jacobi identity is not satisfied in general. Let us consider the map which to every $X \in L$ lets correspond

$$Iz : X \in L \rightarrow Az(X) - Xx$$

where $Az(X) \in \underline{A}z(L)$ and Xx is the value of X at $x = pz$. Let $\underline{G}$ = $Iz(L)$; it is clear that $\underline{G}$ is a vector subspace of $\tilde{G}$. We take up the study of the properties of Iz, thus the structure of $\underline{G}$. First of all Iz is a linear and bijective map of L on $\underline{G}$; in fact the linearity of Iz follows immediately from the definition ; let $AX - Xx$ be an element of $\underline{G}$, we say that to it corresponds at most one $X \in L$ such that

$$Iz(X) = Az(X) - Xx\,, \quad x = pz$$

If Y is another element of L to which corresponds $Az(X) - Xx$, we must have

$$Iz(Y) = Az(X) - Xx \qquad\qquad (x = pz)$$

From the definition of $Iz(Y)$ we then have

$$Az(Y) - Yx = Az(X) - Xx, \qquad\qquad (x = pz)$$

whence

$$Az(Y) = Az(X), \qquad Yx = Xx \qquad (x=pz)$$

In virtue of the above lemma, X and Y coincide on the entire M. Thus Iz is a linear and bijective map. Let us show that

Iz is a homomorphism of L on G. Since it is linear it suffices to show that it preserves the bracket . Let X and Y be two elements of L, we have

$$(14.3) \qquad Iz([X, Y]) = Az\,[X, Y] - [X, Y]x \qquad (x = pz)$$

On the other hand, the last term of the right hand side can be easily put in the form

$$[X, Y]x = Az(X)Yx \ - Az(Y)\,Xx + \Sigma(\widetilde{X}, \widetilde{Y})$$

On putting this expression in (14.3) and on substituting the value of Az[X, Y] taken from (14.2), we obtain

$$Iz([X, Y])$$
$$= [Az(X), Az(Y)] - Az(X)Yx + Az(Y)Xx - \Sigma(\widetilde{X}, \widetilde{Y}) - \Omega(\widetilde{X}, \widetilde{Y})$$

In conformity with (14.2) we then have

$$Iz([X, Y]) = [Az(X) - Xx, Az(Y) - Yx] = [Iz(X), Iz(Y)],$$

where x = pz

Thus Iz defines an isomorphism of L with the Lie-subalgebra G of $\widetilde{G}$. So we have the

Theorem. *Let L be the basic Lie algebra of affine infinitesimal transformations and* $\widetilde{G}$ *the vector space defined by the direct sum*

$$\widetilde{G} = Az(L) + Tpz$$

where Tpz is the tangent vector space to M at x = pz and Az(L) the Lie algebra of endomorphisms of Tpz corresponding to the elements of L. Let suppose that $\widetilde{G}$ *is endowed with the composition law defined by (14.2); Then L is isomorphic to the Lie algebra G = Iz(L)* $\subset \widetilde{G}$ *by the map defined by [1]*

$$Iz\colon X \in L \rightarrow Az(X) - Xx$$

15 The case of Finslerian manifolds

In this section we suppose that M is given the structure of a Finslerian manifold. Let g be the metric tensor of this manifold and ω the corresponding Finslerian connection. We know that the torsion tensor S of this connection is zero; if in the formula (3.1), ω represents the Finslerian connection, then the linear connection $\overline{\omega}$ associated to ω coincides with ω, the matrix L as well its inverse M reduce to identity (local coordinates) and the endomorphism A(X) defined by (6.2) can be written

$$(A(X))^i_j = -(\nabla_j X^i + T^i_{jh} \nabla_0 X^h) \tag{15.1}$$

where ∇ represents the covariant derivative of the Finslerian connection. The results obtained in the preceding paragraphs relative to affine infinitesimal transformations of a general regular linear connection are valid for the Finslerian connection. Let X be an infinitesimal transformation of M. In virtue of (6.6) and (6.7) the Lie derivative of the metric tensor g by X has the form

$$L(X)g_{ij} = -(A(X)_{ij} + A(X)_{ji}) \tag{15.2}$$

where

$$A(X)_{ij} = -(\nabla_j X_i + T_{ijh} \nabla_0 X^h) \tag{15.3}$$

Let us put

$$t(X)_{ij} = -(A(X)_{ij} + A(X)_{ji}) \tag{15.4}$$

For the infinitesimal transformation X we have

$$\nabla_k t(X)_{ij} = g_{ir} L(X) \dot{\Gamma}^r_{jk} + g_{rj} L(X) \dot{\Gamma}^r_{ik} + 2 T_{ijr} L(X) \dot{\Gamma}^r_{ok} \tag{15.5}$$

$$\nabla_{\dot{k}} t(X)_{ij} = g_{ir} L(X) T^r_{jk} + g_{rj} L(X) T^r_{ik} \tag{15.6}$$

From the above formulas we obtain

$$L(X) \dot{\Gamma}^i_{jk} = \frac{1}{2} g^{ih} [\nabla_j t(X)_{hk} + \nabla_k t(X)_{jh} - \nabla_h t(X)_{jk}] - (T^i_{jr} L(X) \dot{\Gamma}^r_{ok} + T^i_{kr} L(X) \dot{\Gamma}^r_{oj} - T_{jkr} g^{ih} L(X) \dot{\Gamma}^r_{oh}) \tag{15.7}$$

(15.8) $$L(X)\, T^i_{jk} = \frac{1}{2} g^{ih} [\nabla_{\dot{j}}\, t(X)_{hk} + \nabla_{\dot{k}}\, t(X)_{jh} - \nabla_{\dot{h}}\, t(X)_{jk}]$$

If X defines an affine infinitesimal transformation (respectively partial) for the Finslerian connection , after (15.5) and (15.6) it is clear that the covariant derivative of two types ∇_k and $\nabla_{\dot{k}}$ (respectively of the type ∇_k) of the tensor t(X) is zero. Conversely let us suppose that covariant derivative of two types ∇_k and $\nabla_{\dot{k}}$ (respectively of type ∇_k) vanishes, we show that infinitesimal transformation X defines an affine infinitesimal transformation (respectively partial) for the Finslerian connection. In fact, if the covariant derivative of type ∇_k of the tensor t(X) vanishes, the relation (15.7) becomes

(15.9) $$L(X)\, \dot{\Gamma}'^{\,i}_{jk} = -\,(T^i_{jr}\, L(X)\, \dot{\Gamma}'^{\,r}_{ok} + T^i_{kr}\, L(X)\, \dot{\Gamma}'^{\,r}_{oj} - T_{jkr} g^{ih}\, L(X)\, \dot{\Gamma}'^{\,r}_{oh})$$

On multiplying the two sides by v^j and taking into account the homogeneity of the torsion tensor we get

(15.10) $$L(X)\, \dot{\Gamma}'^{\,i}_{0k} = -\, T^i_{kr}\, L(X)\, \dot{\Gamma}'^{\,r}_{oo}$$

On multiplying this relation by v^k we get

(15.11) $$L(X)\, \dot{\Gamma}'^{\,i}_{oo} = 0$$

From (15.10) we get

$$L(X)\, \dot{\Gamma}'^{\,i}_{ok} = 0$$

Thus the relation (15.9) gives us

$$L(X)\, \dot{\Gamma}'^{\,i}_{jk} = 0$$

where X is a partial affine infinitesimal transformation. If, in addition, the covariant derivative of type $\nabla_{\dot{k}}$ of the tensor t(X) vanishes from (15.8) we then obtain

$$L(X)\, T^{i}_{jk} = 0$$

Thus X defines an affine infinitesimal transformation for the Finslerian connection. So get the theorem

Theorem. *Given a Finslerian manifold M, in order that an infinitesimal transformation X on M defines an affine infinitesimal transformation (respectively partial) it is necessary and sufficient that the covariant derivative of two types* ∇_k *and* $\dot{\nabla}_k$ *(respectively of the type* ∇_k*) of the tensor* $t(X)_{ij}$ *defined by (15.4) vanishes [1].*

If X defines a partial affine infinitesimal transformation for the Finslerian connection of [(8.6), chap II] we obtain

$$(15.12) \qquad L(X)\, P^{i}_{okl} = -\, v^{j}\, \delta_{l}^{\bullet}\, L(X)\, \dot{\Gamma}^{i}_{jk} = -\, L(X)\, \nabla_{o}\, T^{i}_{kl} = 0$$

16. Case of Infinitesimal Isometries

An infinitesimal isometry is evidently an affine infinitesimal transformation for the Finslerian connection. So the results of the preceding paragraphs apply to the Lie algebras of infinitesimal isometries. Let L be the basic Lie algebra of infinitesimal isometries. Let $\tilde{L}$ be its lift to V(M). To every vector field X $\in$ L we can associate a skew-symmetric endomorphism A(X) of the Euclidean vector space Tpz. For the usual bracket operation these endomophisms define a Lie algebra $\underline{A}$z(L) which is the Lie algebra of connected group Kz(L) of rotations of Tpz. On $\underline{A}$z(L) let us introduce the scalar product defined by

$$(\alpha(z), \beta(z)) = \left(\frac{1}{2}\right)\alpha_{ij}\, \beta^{ij}, \qquad \alpha(z), \beta(z) \in \underline{A}z(L)$$

Let $\underline{\sigma}$z be the Lie algebra of the restricted holonomy group at z and Bz the ortho-complement of $\underline{\sigma}$(z) in the space A of skew-symmetric endomorphisms of Tpz (Bz is complementary subspace of $\underline{\sigma}$(z) in A orthogonal to $\underline{\sigma}$(z) in terms of the above scalar product) Let $\overline{U}$ be a neighbourhood of z $\in$ V(M), if for all z $\in$

$\overline{U}$, $S_{ij}(z)$ belongs to $\underline{\sigma}(z)(\overline{U})$, (see theorem 1 §11 chapter I [1]), so also $\nabla\ S_{ij}(\widetilde{Z})$ for all local vector fields $\widetilde{Z}(Z, \dot{Z})$ on $\overline{U}$. Consequently if for all $z \in \overline{U}$, $\beta_{ij} \in Bz\overline{U}$, by differentiation of $\frac{1}{2} S_{ij}\ \beta^{ij} = 0$. From this we have

$$\frac{1}{2}\nabla S_{ij}(\widetilde{Z})\,\beta^{ij} + \frac{1}{2} S_{ij}\,\nabla\beta^{ij}(\widetilde{Z}) = 0$$

the first term being zero, the second terms must be so whatever be S_{ij} therefore according $\nabla\beta^{ij}(\widetilde{Z})$ belongs to Bz. Let X be an infinitesimal isometry and Az(X) the corresponding endomorphism. $Az(X) \in \underline{A}z(L)$ can be decomposed according as

$$Az(X) = S(z) + \beta(z), \quad (S(z) \in \underline{\sigma}z, , \beta(z) \in Bz) \tag{16.1}$$

On the other hand after (10;4) we have

$$\nabla A(X)_{ji}(\widetilde{Z}) = \Omega_{ij}(\widetilde{X}, \widetilde{Z})$$

From (16.3) it follows

$$-\ \Omega_{ij}(\widetilde{X}, \widetilde{Z}) = \nabla S_{ij}(\widetilde{Z}) + \nabla\beta_{ij}(\widetilde{Z})$$

The left hand side as well as the first term of the right hand side belong to Bz, whence whatever be $\widetilde{Z}$ we have

$$\nabla\beta_{ij}(\widetilde{Z}) = 0$$

Hence it follows that the covariant derivative of the form β is zero. So it is invariant under the holonomy group ψz, in particular we have

$$[\underline{\sigma}(z), \beta(z)] = 0$$

Thus β(z) belongs to connected centralizer $C_0(\sigma z)$ of the restricted holonomy group σz in the group of rotations of Tpz. After (16.3) it follows that Kz(L) is contained in the connected normalizer $N_0(\sigma z)$ of the restricted holonomy group σz in the group of rotations of Tpz. If the Finslerian manifold does not admit 2-form with zero covariant derivative, the group Kz(L) is contained in σz ; if in addition the Lie algebra of infinitesimal isometries is transitive on V(M) then the group Kz(L) coincides with σz, so we get the theorem.

Theorem. *Let M be a Finslerian manifold and V(M) the manifold of non-zero tangent vectors to M Let L be the Lie algebra on M of infinitesimal isometries and* $\widetilde{L}$ *its lift on V(M). If the Finslerian manifold does not have 2-forms with vanishing covariant derivative , the group Kz(L) is contained in the restricted holonomy group σz; if in addition* $\widetilde{L}$ *is transitive on V(M), Kz(L) coincides with σz [1].*

CHAPTER IV

GEOMETRY OF GENERALIZED EINSTEIN MANIFOLDS

(abstract). Let (M, g) be a compact Finslerian manifold of dimension n, and let W(M) be the fibre bundle of unitary tangent vectors to M. Let $\varphi(x, v)$ be a function on W (M). We now introduce the Laplacian $\Delta\varphi = \Delta\varphi(x, v)$ of $\varphi(x, v)$. Now $\Delta\varphi$ decomposes into two parts, $\Delta\varphi = \overline{\Delta}\varphi + \dot{\Delta}\varphi$. We call $\overline{\Delta}\varphi$ the horizontal Laplacian, and $\dot{\Delta}\varphi$ the vertical Laplacian. To the function $\varphi(x, v)$ on W(M) we associate a symmetric 2-tensor $A_{ij}(\varphi)$, and we establish a formula linking the square of $A_{ij}(\varphi)$ to $\overline{\Delta}\varphi$ and a quadratic form $\Phi(\varphi^*, \dot{\varphi})$, where φ^* is the horizontal derivative and $\dot{\varphi}$ the vertical derivative of φ, with coefficients that depend on the curvature of the Finslerian connection. Imposing a certain condition on the curvature tensor, and in case the vertical Laplacian vanishes, we obtain an estimate for the function λ of $\overline{\Delta}\varphi$ ($\overline{\Delta}\varphi = \lambda\varphi$). More precisely, λ cannot always be between zero and n.k where k is a constant > 0. In case M is simply connected and $\lambda = n.k$, then (M, g) is homeomorphic to an n-sphere. This is a generalization of a theorem of Lichnerowicz-Obata in the Riemannian case. The rest of the work is devoted to the deformation of a Finslerian metric. Let $F°(M, g_t)$ be a deformation of (M, g) preserving the volume of W(M). We prove that the critical points $g_0 \in F°(g_t)$ of the integral $I(g_t)$ on W(M) of a certain Finslerian scalar curvature define a Generalized Einstein manifold [6]. We evaluate the second variational of $I(g_t)$ at the critical point g_0 and show that in certain cases and for an infinitesimal conformal deformation, we have $I''(g_0) \geq 0$.

I. Comparison Theorem

1. The Laplacian defined on the Unitary Tangent Fibre Bundle and the Finslerian Curvature

A. Let φ be a differentiable function on W(M). With respect to a local chart of W(M) its derivative is

$$(1.1)\quad d\varphi = D_i\varphi\, dx^i + \partial_i\varphi.\, Dv^i, \qquad D_i\varphi = \delta_i\varphi - \overset{*}{\Gamma}{}^r_{oi}\,\partial_r\varphi = \varphi_i$$

where D_i is the covariant derivation in the Berwald connection. Then by (7.6), (7.9) ch. III)the Laplacian of φ on W(M) decomposes

(1.2) $$\Delta\varphi = \overline{\Delta}\,\varphi + \dot{\Delta}\,\varphi$$

(1.3) $$\overline{\Delta}\,\varphi = -g^{ij} D_i D_j \varphi, \quad \dot{\Delta}\,\varphi = -F^2 g^{ij}\,\dot{\partial}_i\,\dot{\partial}_j\,\varphi, \quad (\dot{\partial}_i = \frac{\partial}{\partial v^i})$$

We call $\overline{\Delta}$ is the horizontal Laplacian and $\dot{\Delta}$ the vertical Laplacian.

Let us consider the symmetric tensor A_{ij} defined by

(1.4) $$A_{ij}(\varphi) = D_i\varphi_j + D_j\varphi_i + \frac{2}{n}\rho g_{ij}, \quad (\varphi_i = D_i\varphi)$$

We choose ρ in such a way that the trace of A vanishes.

(1.5) $$g^{ij}A_{ij} = 2(g^{ij} D_i\,\varphi_j + \rho) = 0, \quad \rho = \overline{\Delta}\,\varphi$$

Let us put

$$(A, A) = \frac{1}{2} A^{ij} A_{ij},$$

Lemma 1. *Let (M, g) be a Finslerian manifold of dimension n; we have*

(1.6) $$\frac{1}{2}(A, A) = (1 - \frac{1}{n})(\overline{\Delta}\,\varphi\,,\,\overline{\Delta}\,\varphi\,) - \Phi(\varphi_*\,\dot{\varphi}\,) + \text{Div on } W(M)$$

where $\Phi(\varphi_*\,\dot{\varphi})$ is a quadratic form in φ_i and $\dot{\varphi}_i = \dot{\partial}_i\varphi$ defined by

(1.7) $$\Phi(\varphi_*, \dot{\varphi}\,) = H_*\,(\varphi_*,\varphi_*) - g^{jl}\,D_j\,H^{\,i}_{olk}\,\varphi^k\,\dot{\varphi}_i - \frac{3}{4}H^i_o{}^{kl}H^{\,j}_{okl}\,\dot{\varphi}_i\dot{\varphi}_j$$

where

(1.8) $$H_*\,(\varphi_*,\varphi_*) = g^{jl}H_{ijkl}\,\varphi^i\varphi^k$$

Proof. We have

$$(1.9) \qquad (A, A) = \frac{1}{2} A^{ij} A_{ij} = g^{ik} g^{jl} D_i \varphi_j A_{kl}$$
$$= g^{ik} D_i (A_{kl} \varphi^l) - g^{ik} \varphi^l D_i A_{kl} - 2 D_o T_j^{kl} \varphi^j A_{kl}$$

Now

$$(1.10) \qquad D_i A_{kl} = D_i D_k \varphi_l + D_i D_l \varphi_k + \frac{2}{n} D_i \rho g_{kl} - \frac{4}{n} \rho D_o T_{ikl}$$

whence

$$(1.11) \qquad -\varphi^l g^{ik} D_i A_{kl} = -\varphi^l g^{ik} D_i (D_k \varphi_l - D_l \varphi_k)$$
$$- \frac{2}{n} \varphi^i D_i \rho + \frac{4}{n} \rho \varphi^i D_o T_i - 2 \varphi^l g^{ik} D_i D_l \varphi_k$$

First, the first term of the right hand side becomes

(1.12)
$$-\varphi^l g^{ik} D_i (D_k \varphi_l - D_l \varphi_k) = -g^{ik} D_i [\varphi^l (D_k \varphi_l - D_l \varphi_k)]$$
$$+ g^{ik} g^{jl} D_i \varphi_j (D_k \varphi_l - D_l \varphi_k)$$
$$= \frac{1}{2} \dot{\varphi}_r H^r{}_{oij} \dot{\varphi}_s H^s{}_o{}^{ij} + \text{Div on } W(M)$$

On the other hand, on using the Ricci identity (§9, chap. I), the last term of (1.11) becomes :

$$(1.13) \quad -2 \varphi^l g^{ik} D_i D_l \varphi_k = 2 \varphi^i D_i \rho + 2 \varphi^j D_o T^{ik}_j A_{ik} - \frac{4}{n} \rho \varphi^i D_o T_i$$
$$-2 H^r{}_{kli} g^{ik} \varphi^l \varphi_r - 2 \partial_{\dot{r}} \varphi_k H^r{}_{oli} \varphi^l g^{ik}$$

Now the last term of the right hand side takes the form

$$(1.14) \quad -2 \partial_{\dot{r}} \varphi_k H^r{}_{oli} \varphi^l g^{ik} = -2 D_k \dot{\varphi}_r H^r{}_{oli} \varphi^l g^{ik}$$
$$= \text{Div on } W(M) + 2 g^{ik} D_k H^r{}_{oli} \varphi^l \dot{\varphi}_r + \dot{\varphi}_r H^r{}_o{}^{kl} \dot{\varphi}_s H^s{}_{okl}$$

Finally let us denote

(1.15)

$$\varphi^i D_i \rho = g^{ij} D_i (\varphi_j \rho) - g^{ij} D_i \varphi_j \rho = \text{Div on } W(M) + (\bar{\Delta}\varphi, \bar{\Delta}\varphi)$$

On taking into account (1.15), (1.14), (1.13) and (1.12) the right hand side of (1.11) is completely calculated, and on putting the result thus obtained in (1.9) and on dividing by 2, we obtain the lemma.

B. Let us consider a 1-form $\zeta = \zeta_i(x)\, dx^i$ on W(M). We have
Lemma 2. *Let (M, g) be a Finslerian manifold of dimension n. On W(M) we have*

(1.16) $$H_*(\zeta, \zeta) = n\,(H(\zeta, u)u, \zeta) + \text{Div on } W(M)$$

Proof. First let us write a Bianchi identity containing the vertical derivation of H (see chapter I formula (10.5) adapted to the Berwald connection)

(1.17) $$\dot{\partial}_m H^i{}_{jkl} + D_l G^i{}_{jkm} - D_k G^i{}_{jlm} = 0$$

where H and G are the two curvature tensors of the Berwald connection . Whence

(1.18) $$V^j \dot{\partial}_m H^i{}_{jkl} = 0$$

Let us consider the 1-form belonging to $\Lambda_1(W)$ defined by its components

$$Z_i = F^{-1} \zeta^j \zeta_r H^r{}_{oji}$$

whence, on taking into account (1.18),

(1.19)

$$F g^{ik} \dot{\partial}_k Z_i = -2 T^{jm}{}_k \zeta_m \zeta_r H^r{}_{oji} g^{ik} + H_*(\zeta, \zeta) - (H(\zeta, u)u, \zeta)$$

Now the first term of the right hand side vanishes, in virtue of the property of the symmetry of the torsion tensor T. On the other hand, let us put

$$\hat{Z}_i = Z_i - u_i \frac{Z_0}{F}, \qquad Z_o = v^r Z_r$$

Now the $\hat{Z}_i$ are the components of a vertical 1-form on W(M). We thus obtain, after (1.19) the formula (1.16).

Lemma 3. *Let f be a differentiable function on M; let us put* $f_i = D_i f$. *We have*

(1.20) $$g^{ij} f_i f_j - n(f_i u^i)^2 = \text{Div on } W(M)$$

This lemma is proved easily, by using the formula (7.6 chap.III).

C. In this paragraph We suppose that the eigenvalue of the Laplacian $\dot{\Delta}$ is zero, φ is therefore independent of the direction. The quadratic form $\Phi(\varphi^*, \dot{\varphi})$ becomes $H_*(\varphi_*, \varphi_*)$ and on identifying ζ with dφ and on using the lemma 2 we obtain

(1.21)
$$\frac{1}{2}(A, A) = (1 - \frac{1}{n})(\Delta\varphi, \Delta\varphi) - n(H(\varphi^*, u)\, u, \varphi^*) + \text{Div on } W(M)$$

where φ* is the gradient associated to dφ. If A = 0, after (1.21) it follows then <H(φ*,u)u, φ*> is positive. We suppose in what follows that (H(φ*u)u, φ*) is positive. Moreover there exists a constant k > 0 such that

(1.22) $$F^{-2} H^i{}_{ojo} \geq k h^i_j \, .$$

where ($h^i_j = \delta^i_j - u^i u_j$, k = constant> 0 and δ the Kronecker symbol.).

Since H_{i0j0} is symmetric ([1] as a consequence of the first Bianchi identity), and positive, we then have

(1.23) $$n(H(\varphi^*, u)u, \varphi^*) \geq n.k[\varphi^i\varphi_i - (\varphi_i u^i)^2], \quad (\varphi_i = D_i\varphi)$$

On taking lemma 3 into account we have :

(1.24) $$n\,(H(\varphi^*, u)u, \varphi^*) \geq (n-1)k\,\varphi^i\varphi_i + \text{Div on } W(M)$$

Now the last term becomes:

$$k\varphi^i\varphi_i = kg^{ij}\,D_j\varphi\,\varphi_i = k.\,(\,\Delta\varphi, \varphi) + \text{Div on } W(M)$$

On putting it in (1.24) we obtain

$$n(H(\varphi^*, u)\,u, \varphi^*) \geq (n-1)k\,(\,\Delta\varphi, \varphi) + \text{Div on } W(M)$$

Thus the right hand side, up to a divergence, is positive. Putting it in (1.21) we obtain by integration on W(M)

$$0 < \frac{1}{2}\langle A, A\rangle \leq \frac{n-1}{n}\,[\langle\Delta\varphi, \Delta\varphi\rangle - nk\,\langle\Delta\varphi, \varphi\rangle]$$

$\dot{\Delta}\varphi$ being zero, φ is a function on M, $\overline{\Delta}\,\varphi$ depends on v in general. We put $\overline{\Delta}\,\varphi = \lambda\,\varphi$ where λ is a function defined on W. We have

$$0 < \frac{1}{2}\langle A, A\rangle \leq \frac{n-1}{n}\int_{W(M)} \lambda(\lambda - nk)\,\varphi^2\eta$$

Now we do not always have

(1.25) $$\lambda \leq nk$$

If $\lambda = nk$, then A = 0. That is to say

(1.26) $$D_i\varphi_j + k\,\varphi\,g_{ij} = 0 \qquad (k = \text{constant} > 0), \quad \varphi_j = D_j\varphi$$

If, in addition, we suppose that M is simply connected, after a result shown in ([3]) the existence of a function φ on M satisfying a differential equation of the second order (1.26) implies that (M, g) is homeomorphic to an n-sphere.

Similarly on supposing that $H_*(\varphi_*, \varphi_*)$ defined by (1.8) is positive and satisfies

$$(1.27) \qquad H_*(\varphi_*, \varphi_*) \geq (n-1)k \left\|\varphi *\right\|^2 \qquad \text{with } k = \text{constant} > 0.$$

On using the formula (1.21) and on reasoning as we have done above we obtain the following result.

Theorem. *Let (M, g) a simply connected compact Finslerian manifold, without boundary, of dimension n. We suppose the curvature tensor H of the Berwald conneciton satisfies the inequality (1.22) or the inequality (1.27) where k is positive constant. Moreover we suppose that the vertical Laplacian of* φ *vanishes. We put* $\overline{\Delta}\varphi = \Delta\varphi = \lambda\varphi$ *Then* λ *cannot always lie between 0 and nk. If* $\lambda = nk$*, (M, g) is then homeomorphic to an n-sphere[6].*

If (M, g) is Riemannian, H_* coincides with Ricci tensor of the Riemannian connection. We have $\lambda \geq nk$. In case of equality, (M, g) is an n-sphere. (see [28] and [33])

2.Case of a manifold with constant sectional curvature

Let us suppose (M, g) is manifold with positive constant sectional curvature.

$$H^i_{jkl} = k(\delta^i_k g_{jl} - \delta^i_l g_{jk}) \quad (k = \text{constant} > 0)$$

And to simplify let k = 1. Then we have

$$H_*(\varphi_*,\varphi_*) = (n-1)\, g^{ij}\, \varphi_i\, \varphi_j = (n-1)\,(\overline{\Delta}\,\varphi, \varphi) + \text{Div on } W(M)$$

$$D_j H^i_{olk} = 0$$

$$-\frac{3}{4} \dot{\varphi}_i H^{i\ kl}_0 \ \dot{\varphi}_j H^j_{okl} = -\frac{3}{2} F^2 \ g^{ij} \ \dot{\varphi}_i \dot{\varphi}_j = -\frac{3}{2} (\dot{\Delta}\varphi, \varphi) + \text{Div on } W(M)$$

The polynomial $\Phi(\varphi_*, \dot{\varphi})$ defined by (1.7) becomes

$$\Phi(\varphi_*, \dot{\varphi}) = (n-1)(\overline{\Delta}\varphi, \varphi) - \frac{3}{2}(\dot{\Delta}\varphi, \varphi) + \text{Div on } W(M)$$

If we denote by λ the function defined by $\overline{\Delta}\varphi = \lambda\varphi$ and by μ the function defined by $\dot{\Delta}\varphi = \mu\varphi$, the formula (1.6) becomes:

$$\frac{1}{2}(A,A) = (\frac{n-1}{n}) \ \sigma(\lambda, \mu) \ \varphi^2 + \text{Div on } W(M)$$

with

$$\sigma(\lambda, \mu) = (\lambda)^2 - n\lambda + \frac{3n}{2(n-1)}\mu$$

Since M is compact and without boundary we have :

$$\frac{1}{2} <A,A> = (\frac{n-1}{n}) \int_{W(M)} \sigma(\lambda, \mu) \ \varphi^2 \ \eta$$

First if $\mu = 0$ and $\lambda = n$ we are in the hypothesis of the preceding theorem. From the polynomial of the second degree σ it follows that if $\mu > \frac{n(n-1)}{6}$, then $\sigma > 0$, and if

$$\mu = \frac{n(n-1)}{6}, \quad \sigma = (\lambda - \frac{n}{2})^2,$$

we have thus an estimate for the eigenvalue of the vertical Laplacian and

$$\frac{n(n-1)}{6} > (n-4).$$

Thus the study of the quadratic form $\Phi(\varphi_*, \dot{\varphi})$ defined by (1.7) turns out to be important for an analysis of the behaviour of the eigenvalues of the Laplacian.

Remark. From (1.6) it immediately follows that if φ is horizontally harmonic, that is to say $\overline{\Delta}\varphi = 0$, then, in case M is compact and without boundary, we have $\varphi_i = 0$ and $A = 0$. If, in addition, M is of constant sectional curvature in the Berwald connection ($K \neq 0$) we have $\dot{\varphi}_i = 0$. Hence φ is constant on W(M). Thus we have the following result:

On a compact Finslerian manifold without boundary and with non-vanishing constant sectional curvature, all functions on W(M), horizontally harmonic, are constant. This result holds true in case the flag curvature is everywhere different from zero [6].

II Deformation of the Finslerian metric. Generalized Einstein Manifolds

1. Fundamental lemma

Let $t \in [-\varepsilon, \varepsilon]$ where ε is sufficiently small > 0. By the deformation of a Finslerian metric we mean a 1-parameter family of this metric $g_{ij}(x, v, t)$. For such a metric $\omega = u_i\, dx^i$, the volume element as well as the connections and curvatures attached to g depend on t. The derivative of the volume element η of W(M) is given by the following lemma proved in ([5]pp. 345-346)

Lemma 1. *The first variational of the volume element of W(M) defined by (§7.chap III) of the fibre bundle of unitary tangent vectors to a Finslerian manifold (M, g) is given by [5]:*

(1.1) $$\eta' = (g^{ij} - \frac{n}{2} u^i u^j)\, g'_{ij}\, \eta, \qquad (u = \frac{v}{F})$$

where the notation ' denotes the derivation with respect to t.
Proof. First of all, we have

$$\omega_t = (\delta F_t/\delta v^i)dx^i, \qquad \omega' = (\delta F'/\delta v^i)dx^i\,, \; (\delta_i^{\bullet} = \partial_i^{\bullet})$$

This derivative commutes with the differentiation d so that from the expression of the volume element ϕ defined in (Chap III, §7) we have

(*) $$\phi' = \omega' \wedge (d\omega)^{n-1} + (n-1)\omega \wedge (d\omega)^{n-2} \wedge d\omega'$$

By a simple calculation from (1.1, Chap II) we obtain

$$\delta F'/\delta v^i = g'_{ir}\, u^r - (F'/F)\, u_i \qquad (u = F^{-1}\, v)$$

$$\omega' = g'_{ir}\, u^r dx^i - (F'/F)\, \omega, \qquad \text{where } \omega = u_i\, dx^i$$

Let us denote by $\theta = \nabla v$, $\beta = \nabla u$, we have

$$\beta^j = F^{-1}(\delta^j{}_k - u^j u_k)\theta^k, \qquad (u_j \beta^j) = 0$$

From the above relation we obtain by differentiation

$$d\omega' = (\partial^2 F'/\delta v^i \partial x^j)\, dx^j \wedge dx^i + F(\delta^2 F'/\delta v^i \delta v^j)\beta^j \wedge dx^i$$

where$(\partial/\partial x^i, \delta/\delta v^j)$denote the Pfaffian derivatives with respect to (dx^j, θ^j) define a basis of $T_z V(M)$ at $z \in V(M)$. The first term of the right hand side is a 2-form in dx, by putting it in (*) it cancels the second term. The coefficients of the second term are given by

$$F(\delta^2 F'/\delta v^i \delta v^j) = g'_{ij} - (F'/F) g_{ij} - g'_{jr} u^r u_i - g'_{ir} u^r u_j + 3\, F'/F u_i u_j$$

Taking into account the above relation the derivative of ϕ can be written

$$(**)\ \phi' = -n\,(F'/F)\phi + g'_{ir} u^r dx^i \wedge (d\omega)^{n-1} + (n-1)\omega \wedge (d\omega)^{n-2} \wedge g'_{ij} \beta^j \wedge dx^i$$

where $$(F'/F) = \frac{1}{2}\, g'_{ij} u^i u^j$$

To evaluate the last two terms of the right hand side of the relation (**) we take an orthonormal frame (e_i) $(i = 1, \ldots n)$ at $x \in M$ such that $u = e_n$, we get $u^n = 1$, $u^\alpha = 0$, $\beta_n = 0$, $\beta_\alpha = \omega_{\alpha n}$ $(\alpha = 1, \ldots n-1)$ where ω_{ij} is the Finslerian connection. Thus the last term of the right hand side of (**) is $g^{i\alpha} g'_{i\alpha}\, \phi$, and the last but one term is equal to $g^{in} g'_{in}\, \phi$, thus their sum is $g^{ij} g'_{ij}\, \phi$. Dividing the two sides of (**) by (n-1)! we get the lemma.

Compact Case

Fundamental Lemma 11. *Let (M, g) be a Finslerian manifold of dimension $n \neq 2$. Let Ψ be a differentiable function, homogeneous of degree zero in v, defined on W(M) and $t_{ij} = (g_{ij})'$ then we have [6]*

(1.2)

$$\Psi \text{ trace } t - n\,\Psi.t(u,u) + \frac{F^2}{(n-2)}(t^{jl} - g^{jl} \text{ trace } t)\dot{\partial}_j\dot{\partial}_l \Psi = \text{Div on } W(M)$$

where $t(u, u) = t_{ij}\, u^i u^j$ and $\dot{\partial}_j = \frac{\partial}{\partial v^j}$

Proof Let $\hat{Y}$ be a co-vector field on W(M) defined by its components

(1.3) $\hat{Y}_j = Y_j - u_j\, Y_o\, F^{-1}$ where $Y_j = F^{-1}\, \psi t_{oj}$, $(\hat{Y}_o = v^j\, \hat{Y}_j = 0)$

where o denotes the contracted multiplication by v. $\hat{Y}$ defines a vertical 1-form; after (7.6, chap III) we have

(1.4) $-\dot{\delta}\,\hat{Y} = F\, g^{jl}\, \dot{\partial}_j\, \hat{Y}_l = \psi \text{ trace } t - n\, \psi.t(u,u) + t^l{}_o\, \dot{\partial}_l \psi$

We are going to calculate the last term of the right hand side. For this, let us consider the co-vector field $\hat{Z}$ defined by

(1.5) $\hat{Z}_k = Z_k - u_k\, Z_o\, F^{-1}$, $Z_k = F\, t^l{}_k\, \dot{\partial}_l \psi$

We have

(1.6)

$$F\, g^{jk}\, \dot{\partial}_j Z_k = t^l{}_o \dot{\partial}_l \psi + F^2 (t^{jl} - \text{trace } t\; g^{jl})\, \dot{\partial}_j \dot{\partial}_l \psi + F^2\, g^{jl}\, \dot{\partial}_j (\text{trace } t\; \dot{\partial}_l \psi)$$

Also,

(1.7) $F g^{jl}\, \dot{\partial}_j\, \hat{Z}_l = F\, g^{jl}\, \dot{\partial}_j\, Z_l - (n-1)\, t^l{}_o\, \dot{\partial}_l\, \psi$

$$= F^2 (t^{jl} - g^{jl} \text{ trace } t)\dot{\partial}_j \dot{\partial}_l \psi - (n-2) t^l{}_o \dot{\partial}_l \psi + F^2\, g^{jl}\, \dot{\partial}_j\, (\text{trace } t\, \dot{\partial}_l \psi)$$

The last term of the right hand side is a vertical divergence after (7.6 chap. III). So also is the left hand side. On putting the expression $t^l{}_o \dot{\partial}_l \psi$ taken from (1.7) in (1.4) we find the formula. Let suppose that (M, g) is compact and without boundary, by integration on W(M) we get

$$(1.8)\quad n\int_{W(M)} \psi t(u,u)\,\eta = \int_{W(M)} [\psi \text{ trace } t + \frac{F^2}{(n-2)}(t^{jl} - g^{jl} \text{ trace } t)\,\partial^{\bullet}_{j}\,{}^{\bullet}_{l}\psi]\eta$$

Let us suppose $\psi = \varphi$ a differentiable function on M, independent of the direction. From (1.8) we have

$$(1.9)\qquad \int_{W(M)} \Phi \text{ trace } t\,\eta = n\int_{W(M)} \Phi\, t(u,u)\,\eta$$

Let us put $\Phi = 1$ and use (1.1) we find

$$(1.10)\qquad (\text{vol } W(M))' = \int_{W(M)} [\text{trace } t - \frac{n}{2} t(u,u)]\eta = \frac{1}{2}\int_{W(M)} \text{trace } t\,\eta = \frac{n}{2}\int_{W(M)} t(u,u)\,\eta$$

2. Variations of scalar curvatures

Let $U(x^i)$ be a local chart of M and $p^{-1}(U)$ (x^i, v^i) the local chart induced on $p^{-1}(U)$. A 1-parameter family of Finslerian connections is represented ([1])

$$(2.1)\qquad \omega^i{}_j|_{p^{-1}(U)} = \overset{*}{\Gamma}{}^i{}_{jk}(x,v,t)\,dx^k + T^i{}_{jk}(x,v,t)\,\nabla v^k$$

where ∇ denotes the Finslerian covariant derivation associated to the 1-parameter family of Finslerian metrics. The coefficients of the Finslerian connection are determined by (5.4 chap II) are written explicitly :

(2.2)

$$\overset{*}{\Gamma}{}^i{}_{jk} = \frac{1}{2} g^{ir}(\delta_k g_{rj} + \delta_j g_{rk} - \delta_r g_{kj}) - (T^i{}_{jr}\overset{*}{\Gamma}{}^r{}_{ok} + T^i{}_{kr}\overset{*}{\Gamma}{}^r{}_{oj} - T_{kjs}\, g^{ir}\overset{*}{\Gamma}{}^s{}_{or})$$

and

$$(2.3)\quad T^i{}_{jk} = \frac{1}{2} g^{ir}\,\delta^{\bullet}_{k}\, g_{rj}, \qquad (\delta_k = \frac{\delta}{\delta x^k},\ \delta^{\bullet}_{k} = \frac{\delta}{\delta v^k})$$

On deriving the relation (2.2) with respect to t we find

(2.4)

$$(\overset{*}{\Gamma}{}^{i}{}_{jk})' = \frac{1}{2} g^{ir} (\nabla_k t_{rj} + \nabla_j t_{rk} - \nabla_r t_{kj})$$

$$- (T^{i}{}_{js} G'{}^{s}{}_{k} + T^{i}{}_{kr} G'{}^{r}{}_{j} - T_{kjs} g^{ir} G'{}^{s}{}_{r})$$

where $t_{ij} = g'_{ij}$, and $G^{i}_{j} = \overset{*}{\Gamma}{}^{i}{}_{oj}$. Let us multiply the two sides of (2.4) by v^j , on taking into account of the property of the torsion tensor we have :

$$(2.5)\quad (\overset{*}{\Gamma}{}^{i}{}_{ok})' = G'{}^{i}{}_{k} = \frac{1}{2} (\nabla_k t^{i}{}_{o} + \nabla_o t^{i}{}_{k} - \nabla^{i} t_{ok}) - 2\, T^{i}{}_{kr} G'{}^{r}$$

Let us multiply this relation by v^k :

$$(2.6)\qquad (\overset{*}{\Gamma}{}^{i}{}_{oo})' = 2\, G'{}^{i} = \nabla_o t^{i}{}_{o} - \frac{1}{2} \nabla^{i} t_{oo}$$

Let now $\pi^{i}{}_{j}$ be the matrix of 1-form of 1-parameter family of Berwald connections associated to g_t defined by :

$$(2.7)\qquad \pi^{i}{}_{j} \big|_{p^{-1}(U)} = G^{i}_{jk} (x, v, t)\, dx^k$$

Let H and G two curvature tensors of this connection. The derivative with respect to t the tensor H becomes (see [5])

$$(2.8)\qquad H'{}^{i}{}_{jkl} = D_k G'{}^{i}{}_{jl} - D_l G'{}^{i}{}_{jk} + G^{i}_{jrk} G'{}^{r}{}_{l} - G^{i}_{jrl} G'{}^{r}{}_{k}$$

where D is the covariant derivative with respect the 1-parmeter family of the Berwald connection. On the other hand the tensor H is related to the curvature tensor R of the Finslerian connection by ([3 p.56]) :

(2.9)

$$R^{i}_{jkl} = H^{i}_{jkl} + T^{i}_{jr} R^{r}_{okl} + \nabla_l \nabla_o T^{i}_{jk} - \nabla_k \nabla_o T^{i}_{jl} + \nabla_o T^{i}_{lr} \nabla_o T^{r}_{jk} - \nabla_o T^{i}_{kr} \nabla_o T^{r}_{jl}$$

Let us denote by $R_{ij} = R^{r}_{irj}$ and $H_{ij} = H^{r}_{irj}$ the corresponding Ricci tensors ; from (2.9) it follows,

(2.10) $$R_{ij}(x, v)\, v^i v^j = H_{ij}\,(x, v)\, v^i v^j = H(v, v)$$

We call the Ricci directional curvature the expression $H(u, u)$ = $R(u, u)$ (see [5] p. 350) we have

Lemma 3. *Let (M, g_t) be a deformation of a Finslerian manifold we have the formula*

(2.11) $$H'(u, u) = H'_{ij}\, u^i u^j = \tau\, t(u, u) + \text{div on } W(M)$$

with

(2.12) $$\tau = g^{ij}\,(D_i D_o T_j + \dot{\partial}_i D_o D_o T_j), \quad t(u, u) = t_{ij} u^i u^j$$

Proof. The derivative of $H(v, v)$ with respect to t is obtained by means of the formula (2.8)

(2.13) $$H'(u, u) = H'_{ij}\, u^i u^j = 2[\nabla_i (F^{-2} G'^{i}) - F^{-2} G'^{i} \nabla_o T_i] - \nabla_o (F^{-2} G'^{i}_{i}) + 4F^{-2} G'^{i} \nabla_o T_i$$

$$= \text{Div on } W(M) + 4\, F^{-2} G'^{i}\, \nabla_o T_i$$

where ∇ denotes the covariant derivative in the Finslerian connection. Let us calculate the last term of the right hand side. In virtue of (2.6) and (7.9 chap III), we have

(2.14) $4 F^{-2} G'^{i} \nabla_o T_i = 2F^{-2} (\nabla_o t^i_o - \frac{1}{2} \nabla^i t_{oo}) \nabla_o T_i$

$$= 2\nabla_o (F^{-2} t^i_o \nabla_o T_i) - 2F^{-2} t^i_o \nabla_o \nabla_o T_i - \{\nabla^i [t(u, u) \nabla_o T_i] - t(u,u) \nabla_o T_i \nabla_o T^i\}$$
$$+ t(u, u)[\nabla^i \nabla_o T_i - \nabla_o T_i \nabla_o T^i]$$
$$= - 2 F^{-2} t^i_o \nabla_o \nabla_o T_i + t(u, u) g^{ij} D_i D_o T_j + \text{Div on } W(M)$$

Now the first term of the right hand side becomes, in virtue of (7.6 Chap III):

$$- 2 F^{-2} t^i_o \nabla_o \nabla_o T_i = F^{-2} g^{ij} \dot{\partial}_i (t_{oo} \nabla_o \nabla_o T_j) + t(u, u) g^{ij} \dot{\partial}_i (D_o D_o T_j)$$

Taking into account this relation and on putting (2.14) in (2.13) one finds the lemma.
Let $\widetilde{H}_{jk}$ be the tensor defined :

(2.15) $$\widetilde{H}_{jk} = \frac{1}{2} (H_{jk} + H_{kj} + v^r \dot{\partial}_j H_{kr})$$

where H_{jk} is the Ricci tensor of the Berwald connection. We have
Lemma 4. *Let (M, g_t) be a deformation of a Finslerian manifold of dimension n : we have*

$$g^{jk} \widetilde{H}'_{jk} = n \tau . t(u, u) + \text{Div on } W(M)$$

where τ is defined by (2.12).
Proof. With the preceding notations we have

$$H'(v, v) = H'_{rs} v^r v^s = F^2 H'_{rs} u^r u^s = F^2 H'(u, u) , (u = \frac{v}{F})$$

Whence by vertical derivation

$$\dot{\partial}_j H'(v, v) = 2 v_j H'(u, u) + F^2 \dot{\partial}_j H'(u, u)$$

A second vertical derivation gives us :

(2.16) $$\widetilde{H}'_{jk} = \frac{1}{2}\, \dot{\partial}_j \dot{\partial}_k H'(v, v) = g_{jk}H'(u, u) + v_j\, \dot{\partial}_k H'(u, u)$$

$$+ v_k\, \dot{\partial}_j H'(u, u) + \frac{1}{2} F^2\, \dot{\partial}_j \dot{\partial}_k H'(u, u)$$

whence

(2.17) $$g^{jk}\, \widetilde{H}'_{jk} = n\, H'(u, u) + \frac{1}{2} F\, g^{jk}\, \dot{\partial}_j\, [F\, \dot{\partial}_k H'(u, u)]$$

The last term of the right hand side is a divergence after (7.6, chapter III). On using the preceding lemma we find the result looked for.

3. Generalized Einstein manifolds

A. In the following we suppose that the Finslerian manifold is compact and without boundary. Let $F(g_t)$ be a 1-parameter family of Finslerian metric. We denote by $\overset{\circ}{F}(g_t)$ the sub-family of Finslerian metrics such that for $t \in [-\varepsilon, \varepsilon]$ the volume of fibre bundle of unitary tangent vectors corresponding to g_t is equal to 1. We look for $g_t \in F^\circ(g_t)$ such that the integral $I(g_t)$ is an extremum.

(3.1) $$I(g_t) = \int_{W(M)} \widetilde{H}_t\, \eta_t \qquad\qquad \widetilde{H}_t = g^{jk}\, \widetilde{H}_{jk}$$

with

(3.2) $$\int_{W(M)} \eta_t = 1$$

where $\widetilde{H}_{jk}$ is defined by (2.15), on taking into account lemma 4, the derivative of $\widetilde{H}_t$ is

(3.3) $$(\widetilde{H}_t)' = (g^{jk}\, \widetilde{H}_{jk})' = -\, t^{jk}\, \widetilde{H}_{jk} + g^{jk}\, \widetilde{H}'_{jk}$$

Thus in virtue of (1.1) the derivative of $\widetilde{H}_t\, \eta_t$ becomes

(3.4)
$(\widetilde{H}_t \eta_t)' = [- t^{jk} \widetilde{H}_{jk} + \widetilde{H}_t t^{jk} g_{jk} + n \psi t(u, u)] \eta + \text{Div on } W(M)$
where

(3.5) $$\psi = \tau - \frac{\widetilde{H}}{2}$$

and τ is defined by (2.12) On replacing in (3.4) the expression n ψ t(u, u) drawn from the formula (1.2) of lemma 11 we obtain finally

(3.6) $$(\widetilde{H}_t \eta_t)' = - \widetilde{A}_{jk} t^{jk} \eta + \text{Div on } W(M)$$

where we have put :
(3.7)
$$\widetilde{A}_{jk} = \widetilde{H}_{jk} - g_{jk} (\psi + \widetilde{H}) + \frac{1}{n-2} F^2 g_{jk} g^{rs} \dot{\partial}_r \dot{\partial}_s \psi - \frac{1}{n-2} F^2 \dot{\partial}_j \dot{\partial}_k \psi$$

M being compact and without boundary , on integrating on W(M) we get

(3.8) $$I'(g_t) = - \int_{W(M)} \widetilde{A}_{jk} t^{jk} \eta_t$$

A point $g_o \in F^\circ(g_t)$ (t= 0) is critical when $I'(g_0)= 0$. So at the critical point g_0 the tensor $\widetilde{A}_{jk}$is L_2 orthogonal to t_{jk}.

Since the volume of W(M) is supposed to equal to one, after (1.10) we have

(3.9)
$$<g, t> = \int_{W(M)} \text{trace } t . \eta = 0 \text{ and } <t, u\otimes u> = \int_{W(M)} t(u, u) \eta = 0$$

Thus the tensors g and u⊗u are L_2 orthogonal to t. If $\widetilde{A}$ is a linear combination of g and u⊗u with constant coefficients at a point $g_o \in F^\circ(g_t)$ (t= 0), that is to say if

(3.9)'. $\widetilde{A}_{jk} = cg_{jk} + bu_j u_k$ where c and b are constants,

then from (3.8), (3.9) and from the above relation it follows that g_0 is a critical point of $I(g_t)$ at t=0. *But* $\widetilde{A}$ *is non-decomposable. We now show that* $b = 0$. *In fact, on deriving vertically the above relation, and taking into account the expression of* $\widetilde{A}$ defined by (3.9)', we obtain

$$g^{ik} v^j \dot{\partial}_i \widetilde{A}_{jk} = \frac{1}{n-2} F^2 g^{jk} \dot{\partial}_j \dot{\partial}_k \psi = (n-1)b, \qquad (n\neq 2)$$

Now the second term is a divergence on W(M). On integrating on W(M) we have b = 0. Thus we have

(3.10)

$$\widetilde{A}_{jk} = \widetilde{H}_{jk} - (\psi + \widetilde{H})g_{jk} + \frac{1}{n-2} F^2 g_{jk}\, \phi - \frac{1}{n-2} F^2 \dot{\partial}_j \dot{\partial}_k \psi = C\, g_{jk}$$

where

(3.11) $\phi = g^{rs} \dot{\partial}_r \dot{\partial}_s \psi, \qquad \psi = \tau - \frac{\widetilde{H}}{2}$

After (3.10) the vertical derivative of Cg_{jk} is completely symmetric, so also the vertical derivative of $\widetilde{A}_{jk}$ that is to say $\dot{\partial}_l \widetilde{A}_{jk}$. Thus on writing the equality $\dot{\partial}_l \widetilde{A}_{jk} = \dot{\partial}_j \widetilde{A}_{kl}$, and on suppressing the common elements of this equality we find

(3.12)

$$-\dot{\partial}_l (\psi + \widetilde{H})g_{jk} + \frac{1}{n-2} 2\, v_l\, g_{jk}\, \Phi + \frac{1}{n-2} F^2 g_{jk}\, \dot{\partial}_l \Phi - \frac{1}{n-2} 2v_l\, \dot{\partial}_j \dot{\partial}_k \psi$$

$$= -\dot{\partial}_j (\psi + \widetilde{H})g_{kl} + \frac{1}{n-2} 2v_j g_{kl} \Phi + \frac{1}{n-2} F^2 g_{kl}\, \dot{\partial}_j \Phi - \frac{1}{n-2} 2v_j \dot{\partial}_k \dot{\partial}_l \psi$$

Let us multiply the two sides of (3.12) by g^{jk} we obtain

$$(3.13)\quad (1-n)\dot{\partial}_l (\psi + \widetilde{H}) + 2\, v_l\, \Phi + \frac{n-1}{n-2} F^2\, \dot{\partial}_l \Phi - \frac{2}{n-2} \dot{\partial}_l \psi = 0$$

where ψ and $\widetilde{H}$ are homogeneous of degree zero in v while Φ is homogeneous of degree -2 in v; on multiplying (3.13) by v^l we obtain

(3.14) $$-\frac{1}{n-2} F^2 \Phi = 0, \quad \Phi = g^{rs} \dot{\partial}_r \dot{\partial}_s \psi = 0 , n \neq 2$$

Let us multiply the relation $\Phi = 0$ by $F^2 \psi$ we get

$$0 = F^2 g^{rs} \dot{\partial}_r (\psi \dot{\partial}_s \psi) - F^2 g^{rs} \dot{\partial}_r \psi \dot{\partial}_s \psi$$

After (7.6 chap III) the first term of the right hand side is a divergence, thus by integration on W(M) we obtain :

(3.15) $$\int_{W(M)} F^2 g^{rs} \dot{\partial}_r \psi \dot{\partial}_s \psi \eta = 0$$

whence

(3.16) $$\dot{\partial}_r \psi = 0$$

So from (3.13) it follows

(3.17) $$\dot{\partial}_l \widetilde{H} = 0$$

On taking into account the expression ψ defined by (3.5) , the relation (3.10) becomes

(3.18) $$\widetilde{H}_{jk} - \left(\tau + \frac{1}{2} \widetilde{H} \right) g_{jk} = C.g_{jk}$$

On multiplying the above relation by g^{jk} and on dividing by n we obtain

(3.19) $$\frac{1}{n} \widetilde{H} - \left(\tau + \frac{1}{2} \widetilde{H}\right) = C$$

From (3.18) and (3.19) it follows

(3.20) $$\widetilde{H}_{jk} = \frac{1}{n}\widetilde{H} g_{jk}$$

where $\widetilde{H}$ is independent of the direction.

Definition . *A Finslerian manifold is called a generalized Einstein manifold (G.E.M) if the Ricci directional curvature is independent of the direction, that is to say ([*5],[6])

(3.21) $$\widetilde{H}_{jk}(x, v) = C(x)\, g_{jk}(x, v)$$

where C(x) is a function defined on M. We have proved the following theorem:

Theorem. *For a compact Finslerian manifold without boundary the Finslerian metric $g_o \in F°(g_t)$ at the critical point ($t = 0$, $g_o = g(0)$) of the integral $I(g_t)$ defines a Generalized Einstein Manifold*[6].

For a Generalized Einstein manifold $\widetilde{H}$ is independent of the direction, and after (3.19) it is defined by $\frac{2n}{2-n}(\tau + C)$. Let us suppose that $\widetilde{H}$ is constant then τ must be constant But τ being defined by (2.12) is a divergence on W(M) . Consequently $\tau = 0$.

We have

Corollary. *If the Finslerian metric $g_o \in F°(g_t)$ at $t = 0$ is critical for the integral $I(g_t)$ and defines at this point a manifold with Ricci directional curvature constant, then we have at this point*

($\widetilde{H}_{jk} = \frac{1}{n}\widetilde{H} g_{jk}$, $\widetilde{H}$ = constant), ,)and

(3.22) $$\tau = g^{ij}(D_i D_o T_j + \dot{\partial}_i D_o D_o T_j) = 0$$

B. Let us consider now the integral

(3.23) $$I_1(g_t) = \int_{W(M)} H_t(u, u)\, \eta_t$$

With the condition that the volume of W(M) is constant and is equal to one we look for a $g_t \in F°(g_t)$ such that $I_1(g_t)$ is an extremum. To derive $I_1(g_t)$ with respect to t we have, first of all,:

(3.24) $$[H_t(u, u)]' = (H_{ij})'u^i u^j + H_{ij}[(u^i)'u^j + u^i(u^j)']$$

where $(u^i)' = -\frac{F'}{F}u^i = -\frac{1}{2}t(u, u)u^i$. In virtue of the lemma 3 the relation (3.24) becomes

$$[H_t(u, u)]' = [\tau - H(u, u)]\, t(u, u) + \text{Div on } W(M)$$

whence

(3.25) $$[H(u, u)\eta_t]' = [\psi.t(u, u) + H(u, u)\,\text{trace } t\,]\eta + \text{Div on } W(M)$$

where we have put

(3.26) $$\psi = \tau - (\frac{n}{2} + 1)\, H(u, u)$$

Substituting in (3..25) the value of ψ t(u, u) drawn from the formula (1.2) we get :

(3.27) $$I'_1(g_t) = \int_{W(M)} \widetilde{B}_{jk}\, t^{jk}\, \eta_t = 2 < B, t >$$

Thus

(3.28) $$\widetilde{B}_{jk} = [\, H(u, u) + \frac{\Psi}{n} - \frac{F^2}{n(n-2)}\Phi\,]g_{jk} + \frac{F^2}{n(n-2)}\partial^{\bullet}_j{}_{;k}\Psi$$

where

(3.29) $$\Phi = g^{jk}\,\partial^{\bullet}_j{}_{;k}\Psi$$

Hence, by identical reasoning as above, at the critical point t = 0, $g_o = g(o) \in F°(g_t)$ the integral $I_1(g_t)$, the tensor $\widetilde{B}_{jk}$ becomes

(3.30)

$$\widetilde{B}_{jk} = [\, H(u,u) + \frac{\Psi}{n} - \frac{F^2}{n(n-2)}\Phi\,]g_{jk} + \frac{F^2}{n(n-2)}\dot{\partial}_j\dot{\partial}_k\psi = c\,g_{jk}$$

In fact, after (3.9), $\widetilde{B}_{jk}$ is proportional to g_{jk} and $u_j \otimes u_k$. As in paragraph A we show that b = 0.

Let us multiply the two sides by g^{jk} and contracting, we have

$$[\, H(u,u) + \frac{\Psi}{n} - \frac{F^2}{n(n-2)}\Phi\,] + \frac{F^2}{n^2(n-2)}\phi = C \tag{3.31}$$

Let us multiply the two sides by g_{jk} and subtracting from (3.30) the relation thus obtained, we have

$$\frac{F^2}{n(n-2)}\dot{\partial}_j\dot{\partial}_k\psi - \frac{F^2}{n^2(n-2)}\Phi g_{jk} = 0$$

whence, on multiplying by v^j and v^k we obtain

$$\Phi = 0, \qquad \dot{\partial}_j\Psi = 0. \tag{3.32}$$

From (3.31) and (3.26) we then have

$$H(u,u) = C - \frac{\Psi}{n} = \frac{2n}{n-2}(c - \frac{1}{n}\tau) \tag{3.33}$$

with τ is independent of the direction. Thus H(u, u) is independent of the direction. And g_o is a generalized Einstein manifold. If on the other hand we suppose that H(u, u) is constant then $\tau = 0$. Thus we obtain the same conclusion. If we write after (3.30) that $\dot{\partial}_l \widetilde{B}_{jk} = \dot{\partial}_j \widetilde{B}_{kl}$ and proceed then in the same manner as the preceding paragraph.

Theorem. *The Finslerian metric $g_o \in F°(g_t)$ at the critical point $(t= 0, g_o=g(0))$ of the integral $I_1(g_t)$ defines a generalized Einstein manifold. Moreover if H(u, u) is constant, then $\tau = 0$*

4. Second variational of the integral I(g$_t$)

A. In view of reducing the calculation of the second derivative of $I(g_{t})$ we are going to suppose that the trace of the torsion is invariant by deformation and prove some lemmas.

Lemma 5. *Let T' be the derivation with respect to $t \in [-\varepsilon, \varepsilon]$ of the torsion tensor of Finslerian connection and $t_{ij} = g'_{ij}$,the following conditions are equivalent*

(4.1) $(T^i_{jk})' = 0 \Leftrightarrow \partial^{\bullet}_j t^i_k = 0$

(4.2) $(T^i_{ji})' = 0 \Leftrightarrow g^{jk}\partial^{\bullet}_j t^i_k = 0 = \partial^{\bullet}_j \text{ trace } t$

Proof : we have

$$\frac{1}{2}\partial^{\bullet}_j g_{ki} = T_{kij} = g_{kr} T^r_{ij}$$

Let us derive the two sides with respect to t

$$\frac{1}{2}\partial^{\bullet}_j g'_{ki} = \frac{1}{2}\partial^{\bullet}_j t_{ki} = t_{kr} T^r_{ij} + g_{kr} (T^r_{ij})' = \frac{1}{2}\partial^{\bullet}_j (g_{ir} t^r_k)$$

$$= t^r_k T_{rij} + \frac{1}{2} g_{ir} \partial^{\bullet}_j t^r_k$$

whence

(4.3) $\frac{1}{2} g_{ir} \partial^{\bullet}_j t^r_k = g_{kr} (T^r_{ij})'$

From this relation we deduce the lemma.

Lemma 6. *Let us suppose that the trace of the torsion is invariant by deformation and Ψ is a function on W(M) homogeneous of degree zero at v; we then have*

(4.4) $-F^2 \text{ trace } t \, g^{jl} \partial^{\bullet}_j \;_l \Psi = \text{Div on } W(M)$

This lemma follows from the formula (7.6 chap III) and (4.2). In fact, trace t is a function on M. Under the condition of the lemma 6 in particular (1.2.II) becomes

(4.5) $\quad n\Psi.t(u, u) = \Psi \text{ trace } t + \dfrac{1}{n-2} F^2 t^{ij} \partial^{\bullet\bullet}_{i\ j} \Psi + \text{Div on } W(M)$

Similarly we have

Lemma 7. *Let (M, g_t)a deformation of a Finslerian manifold we have*

(4.6) $\qquad F t(u, u) g^{ij} \partial^{\bullet}_i [F \partial^{\bullet}_j H'(u, u)]$

$\qquad\qquad = 2 [\text{trace } t - n\, t(u, u)] H'(u, u) + \text{Div on } W(M)$

This lemma is obtained by applying twice the formula (7.6 chap III).

   B. In the preceding paragraphs we have put

(4.7) $\qquad \sigma_{(m,r,l)} \nabla_m R^i_{jrl} + \sigma_{(m,r,l)} (F^2 \dfrac{1}{3} K_m + Kv_m) \nabla_o Q^i_{jrl} = 0$

On supposing that the trace of the torsion is invariant by deformation the first variational of $I(g_t)$ becomes

(4.8) $\qquad\qquad I'(g_t) = - \displaystyle\int_{W(M)} \hat{A}_{ij} t^{ij} \eta_t$

where in virtue of (4.5) the expression of $\widetilde{A}_{ij}$ is defined by (3.7) becomes $\hat{A}_{ij}$

(4.9) $\qquad\qquad \hat{A}_{ij} = \widetilde{H}_{ij} - g_{ij} (\psi + \widetilde{H}) - \dfrac{F^2}{n-2} \partial^{\bullet}_{i\ j} \psi$

where

(4.10) $\qquad\qquad t_{ij} = g'_{ij}$ and $\qquad\qquad \psi = \tau - \widetilde{H}/2.$

At the critical point $t = 0$, $I'(g_o) = 0$, (M, g_o) is a generalized Einstein manifold.

$$\tilde{H}_{ij} = \frac{1}{n}\tilde{H}\, g_{ij}\,,\ \hat{A}_{ij} = C g_{ij} \text{ where C is a constant.}$$

$\tilde{H}$, ψ and τ are independent of the direction and after (3.19) we have at t= 0

(4.11)
$$\tilde{H} = \frac{2n}{2-n}(\tau + C)$$

$\hat{A}_{ij}$ being a symmetric tensor homogeneous of degree zero at v, the derivation under the integral sign of (4.8) is :

(4.12)
$$(t^{ij}\hat{A}_{ij}\eta)' = [-2t^{ik}t^{j}_{k}\hat{A}_{ij} + t^{ij}(\hat{A}_{ij})' + \hat{A}^{ij}t'_{ij} + t^{ij}\hat{A}_{ij}(g^{rs} - \frac{n}{2}u^{r}u^{s})\,t_{rs}]$$

We evaluate this expression at the point t = 0; we have

(4.13)
$$-2t^{ik}\,t^{j}_{k}\,\hat{A}_{ij} = -2c\,t^{ij}\,t_{ij}$$

(4.14)
$$t^{ij}\hat{A}_{ij}\,g^{rs}\,t_{rs} = c(\text{ trace } t\,)^{2}$$

(4.15)
$$-\frac{n}{2}t^{ij}\hat{A}_{ij}\,t(u, u) = -\frac{n}{2}c \text{ trace t. } t(u, u)$$

The term $\hat{A}^{ij}\,t'_{ij} = C\,g^{ij}\,g''_{ij}$. Now the volume W(M) = 1 after (1.10) we have

(4.16)
$$\int_{W(M)} g^{ij}g'_{ij}\eta_{t} = 0$$

By derivation we obtain at the point t = 0

(4.17)
$$\int_{W(M)} \hat{A}^{ij}\,t'_{ij}\,\eta = C\int_{W(M)} \{t^{ij}t_{ij} - \text{trace t [trace t} - \frac{n}{2}\text{ H(u, u)] }\}\eta$$

From (4.12) it remains to calculate the term $t^{ij}(\hat{A}_{ij})'$ at $t = 0$. Now at this point ψ has vertical derivative zero; by (4.9) we then have at this point :

(4.18) $t^{ij}(\hat{A}_{ij})'|_{t=0}=$

$$[t^{ij}\widetilde{H}'_{ij} - (\psi+\widetilde{H})' \operatorname{trace} t - (\psi + \widetilde{H})t^{ij} t_{ij} - \frac{F^2}{(n-2)} t^{ij}\dot{\partial}_i \dot{\partial}_j \psi']_{t=0}$$

From (2.16) we obtain

(4.19)

$$t^{ij} \widetilde{H}'_{ij}=H'(u, u) \operatorname{trace} t + \frac{3}{2} t^i{}_o \dot{\partial}_i H'(u, u) + \frac{1}{2} Ft^{ij}\dot{\partial}_i [F\dot{\partial}_j H'(u, u)]$$

Now the trace of torsion is invariant by deformation, in virtue of (4.2) we obtain :

$$t^i{}_o \dot{\partial}_i H'(u, u)=[n\, t(u, u)-\operatorname{trace} t]H'(u, u) + \text{Div on } W(M)$$

$$F\, t^{ij} \dot{\partial}_i [F \dot{\partial}_j H'(u, u)]$$

$$= (n-1) [n\, t(u, u) - \operatorname{trace} t]H'(u, u) + \text{Div on } W(M)$$

Thus (4.19) becomes

(4.20) $t^{ij}(\widetilde{H}_{ij})'=\frac{1}{2} n[(n+2)t(u, u) - \operatorname{trace} t]H'(u, u) + \text{Div on } W(M)$

For the last term of the right hand side of (4.18) we have, on taking into account (4.2)

(4.21) $F^2 t^{ij}\dot{\partial}_i \dot{\partial}_j \psi' = F^2 g^{ik} \dot{\partial}_i (t^j{}_k\dot{\partial}_j \psi')$

$$= F g^{ik} \dot{\partial}_i (F t^j{}_k \dot{\partial}_j \psi') - t^j{}_o\dot{\partial}_j \psi'$$

$$= (n-2) t^j{}_o \dot{\partial}_j \psi' + \text{Div on } W(M)$$

$$=(n-2) F g^{jk}\dot{\partial}_j (F^{-1}t_{ko}\psi')+(n-2) [t(u, u)-\operatorname{trace} t] \psi'+\text{Div on } W(M)$$

$$= (n-2)[nt(u, u) – \operatorname{trace} t] \psi' + \text{Div on } W(M)$$

At the point $t = 0$ we have

(4.22) $$(\psi + \widetilde{H})_{t=0} = \frac{1}{n}\widetilde{H} - C$$

Now $\psi = \tau - \frac{1}{2}\widetilde{H}$, we have after (4.21)

(4.23) $$-\frac{1}{n-2}F^2 t^{ij} \dot{\partial}_i \dot{\partial}_j \psi' - (\psi + \widetilde{H})' \text{ trace } t$$

$$= -n\, t(u, u)\tau' + \frac{1}{2} n t(u, u)\widetilde{H}' - \text{trace } t\, \widetilde{H}' + \text{div on } W(M)$$

On taking into account (2.17), $\widetilde{H}'$ becomes at the point t = 0

(4.24) $$\widetilde{H}' = -\frac{1}{n}\widetilde{H} \text{ trace } t + n H'(u, u) + \frac{1}{2} F g^{ij} \dot{\partial}_i [F \dot{\partial}_j H'(u, u)]$$

On multiplying the two sides by $\frac{1}{2}$n t(u, u) and on using the formula (4.6) of the lemma 7 we obtain

(4.25) $$\frac{1}{2} n\, t(u, u)\, \widetilde{H}' = -\frac{1}{2}\widetilde{H} \text{ trace } t \,.\, t(u, u)$$
$$+ \frac{1}{2} n \text{ trace } t\, H'(u, u) + \text{Div on } W(M)$$

Similarly, on multiplying the two sides of $\widetilde{H}'$ by - trace t from the fact that the trace of the torsion remains invariant by deformation then trace t is a function on M, as well as the last term of the right hand side of (4.24) will be a divergence; we have

(4.26) $$-\text{trace } t\, \widetilde{H}' = \frac{1}{n}\widetilde{H} \text{ trace } t - n \text{ trace } t\, H'(u, u) + \text{Div on } W(M)$$

On adding (4.25) to (4.26) and on putting them in (4.23) we obtain

(4.27)

$$-\frac{1}{n-2}F^2\, t^{ij}\, \partial_{\dot{i}}{}_{\dot{j}}\,\psi' - (\psi + \widetilde{H})'\ \text{trace}\ t = -n\, t(u,u)\tau'$$

$$+\frac{1}{n}\widetilde{H}\,(\text{trace}\ t)^2 - \frac{1}{2}\widetilde{H}\ \text{trace}\ t\ t(u,u) - \frac{1}{2}n\,\text{trace}\ t H'(u,u) + \text{Div on } W(M)$$

From (4.22) we have

$$(4.28) \qquad -(\psi + \widetilde{H})_{t=0}\, t^{ij} t_{ij} = (C - \frac{1}{n}\widetilde{H})\, t^{ij} t_{ij}$$

Thus on adding side by side the relations (4.20) and (4.27) and (4.28) we get the expression $t^{ij}(\hat{A}_{ij})'$ at the point t = 0 then on adding the result thus obtained to (4.13) , (4.14) (4.15) and (4.17) we obtain finally :

$$(4.29) \quad I''(g_o) = \int_{W(M)} \{\frac{1}{n}\widetilde{H}\,[(2(t,t) - (\text{trace}\ t)^2 + \frac{n}{2}\text{trace}\ t.\ t(u,u)\,]$$

$$-\frac{n}{2}[(n+2)t(u,u) - 2\ \text{trace}\ t]\ H'(u,u) + n\ t(u,u)\ \tau'\}_{t=0}\eta$$

where

$$\tau = g^{ij}\,(D_i D_o T_j + \partial_{\dot{i}} D_o D_o T_j), \qquad (t,t) = \frac{1}{2}t^{ij}\, t_{ij}$$

(4.30)

$$H'(u,u) = H'_{ij}u^i u^j = F^{-2}\,\nabla_i\,(\nabla_o t'^i{}_o - \nabla^i t_{oo}) + F^{-2}\,(\nabla_o t'^i{}_o - \nabla^i t_{oo})\,\nabla_o T_i$$

$$- F^{-2}\nabla_o\,[\frac{1}{2}\,(\nabla_o t'^i{}_i + T_i\,\nabla^i t_{oo}) - T_i \nabla_o t'^i{}_o\,]$$

Formula of the second variational.

Theorem. *Let (M, g_t) be a deformation of a compact Finslerian manifold without boundary which leaves invariant the torsion trace. The second derivative of the integral I(g_t) defined by (3.1) at the critical point (t = 0, g_t = g_o) is given by the formula (4.29)*

5. Case of a Conformal Infinitesimal Deformation.

Let us suppose (M, g_t) be a conformal infinitesimal deformation. Then we have

$$t_{ij} = g'_{ij} = 2\,\varphi(x)\,g_{ij} \tag{5.1}$$

where φ is a differentiable function defined on M. We suppose in what follows that τ is everywhere zero. Therefore $\widetilde{H}$ is a constant. Also after the lemma 5 it follows that torsion tensor is invariant by infinitesimal conformal deformation.
We are going to evaluate the expression under the integral sign. The term containing $\widetilde{H}$ in $I''(g_t)$ is

$$2\,(2 - n)\,\varphi^2\,\widetilde{H} \tag{5.2}$$

For the term containing $H'(u, u)$ we use the expressions $G'^{\,i}$ and $G'^{\,i}_{\;j}$ determined by (2.5) and (2.6) and for a conformal infinitesimal deformation

(5.3)

$$-n(n-2)F^{-2}\,\varphi D_o\varphi_o - n(n-2)\,\varphi[g^{ij}\,D_i\,\varphi_j + D_o\,(T^i\varphi_i) + 2\,\varphi^i D_o\,T_i]$$

In virtue of lemma 3 we have

$$F^{-2}\,D_o\,\varphi_o\varphi = -\left(\frac{\varphi_o}{F}\right)^2 + \mathrm{Div} = -\frac{1}{n}\varphi^i\varphi_i + \mathrm{Div} \tag{5.4}$$

$$= \frac{1}{n}\varphi g^{ij}\,D_i\,\varphi_j + \mathrm{Div}\ \text{on } W(M)$$

Now $\tau|_{t=0}$ is zero, therefore we have

$$\varphi\varphi^i D_o\,T_i = \frac{1}{2}g^{ij}D_j\,(\varphi^2 D_o\,T_i) - \frac{1}{2}\varphi^2 g^{ij}\,D_j\,D_oT_i \tag{5.5}$$

$$= \mathrm{Div}\ \text{on } W(M)$$

It remains to evaluate the term $\varphi\ D_o(T^i\varphi_i)$ at the same time as the term $2n\tau'|_{t=0}$. For this last on making explicit the formulas (2.5) and (2.6) for a conformal infinitesimal deformation we have the torsion trace being invariant

$$(5.6) \qquad (D_0\ T_j)' = F^2\ \nabla_{\dot{i}}\ T_j\ \varphi^i + v_j\ \varphi^i T_i + \varphi_o T_j$$

On the other hand $\tau\,|_{t=0} = 0$, the derivative of τ becomes:

$$\tau'|_{t=0} = [g^{ij}\ (D_i D_o T_j)' + g^{ij}\ \partial_{\dot{i}}\ (D_o D_o T_j)'\ ]$$

whence on multiplying by $2n\ \varphi(x)$

$$2n\varphi\ (x)\ \tau'|_{t=0} = 2n\ \varphi g^{ij}\ (D_i D_o T_j)' + \text{Div on } W(M)$$

On using (5.5) and (5.6) we obtain , after simplifying

$$2n\varphi(x)\tau'|_{t=0} = -2n\ [F^2\ \nabla_{\dot{i}}\ T_j\ \varphi^i\varphi^j + 2\ T^i\ \varphi_i\varphi_o] + \text{Div on } W(M)$$

Thus the expression under the integral sign is

(5.7)

$$2(n-1)(n-2)\ (\Delta\varphi - \frac{1}{n-1}\tilde{H}\ \varphi,\ \varphi) - 2n\ F^2\ \nabla_{\dot{i}}\ T_j\ \varphi^i\varphi^j + n(n-6)\ T^i\varphi_i\varphi_o + \text{Div on } W(M)$$

On direct calculation we obtain

$$F^2\ \nabla_{\dot{i}}\ T_j\ \varphi^i\varphi^j = (n-2)\ T^j\ \varphi_j\varphi_o + F^2\ \left\|\nabla_{\dot{i}}\varphi_j\right\|^2 - F^2\ Q_{ij}\varphi^i\varphi^j + \text{Div on } W(M)$$

And the expression $T^j\varphi_j\varphi_o$ is evaluated , with the help of lemma 5, Chapter VIII as a function $F^2\ Q_{ij}\ \varphi^i\varphi^j$. Thus finally we have the expression

$$(5.8)\quad I''(g_o) = \int_{W(M)} [2(n-1)(n-2)\,(\Delta\varphi - \frac{1}{n-1}\widetilde{H}\varphi,\ \varphi) - \frac{8n}{(n-2)} Q_{ij}\,\varphi^i\varphi^j - 2n\,F^2 \left\|\nabla_i\varphi_j\right\|^2]\eta$$

We thus obtain the formula for the second variational for an infinitesimal conformal deformation. After a formula established in ([2] page 224) we have

$$(n-2)\,F^2 \left\|\nabla_i\varphi_j\right\|^2 + (n+2)\,F^2 Q_{ij}\,\varphi^i\varphi^j = \text{Div on } W(M) + (n-2)\,\varphi^i\,(D_0 T_i)'$$

On using $\tau = 0$, and the relations (2.5, II), (2.6, II) and (7.6 and 7.9 chap III) we prove that the last term of the right hand side of the above relation is a divergence. In this case we have

$$(5.9)\qquad (n-2)\,F^2 \left\|\nabla_i\varphi_j\right\|^2 + (n+2)\,F^2 Q_{ij}\,\varphi^i\varphi^j = \text{Div on } W(M)$$

On the other hand, on taking into account the lemmas 5 and 6 of Chapter VIII, the relation (5.15 Chapter VIII) becomes

$$(5.10)\ (\Delta\varphi,\ \varphi) = \frac{1}{n-1}\widetilde{H}\varphi^2 - \frac{n}{(n-1)(n-2)} F^2 Q_{ij}\,\varphi^i\varphi^j + \text{Div on } W(M)$$

Taking into account the relation (5.9), (5.10) becomes

$$(5.11)\ (\Delta\varphi - \frac{1}{n-1}\widetilde{H}\varphi\,,\ \varphi) = \frac{n}{(n-1)(n+2)}\,.F^2 \left\|\nabla_i\varphi_j\right\|^2 + \text{Div } W(M),$$

From this we deduce, by integrating over W(M),

$$(5.12)\qquad < \Delta\varphi - \frac{1}{n-1}\widetilde{H}\varphi\,,\ \varphi > \ \geq 0$$

If the integral

$$(5.13) \qquad \int_{W(M)} F^2 Q_{ij}\, \varphi^i \varphi^j \;\eta = 0,$$

on taking into account the above relations and (5.12) and (5.8), we have

$$(5.14) \qquad I''(g_o) \geq 0.$$

Theorem. *Let (M, g) be a compact Finslerian manifold without boundary ($n \neq 2$). We suppose that τ is everywhere zero, $\tau' \big|_{t=0} = 0$ and the Ricci vertical curvature Q_{ij} satisfies (5.13). Then at the critical point $g_o \in F^{\circ}(g_t)$, ($t=0$) of the integral $I(g_t)$ defined by (3.1) and for a conformal infinitesimal deformation, the second derivation is positve[6].*

Remark. Let λ be a function such that $\Delta\varphi = \lambda\varphi$ and λ_1 the least value of λ at the point y $\in$W. Let us put $\lambda_1 = \min_{y \in W} \lambda_1(y)$. Let us suppose $\widetilde{H}$ to be a positive constant. From the relation (5.12), it follows that

$$\lambda_1 \geq \frac{1}{n-1} \widetilde{H}$$

CHAPTER V

Properties of Compact Finslerian Manifolds of Non-negative Curvature

(Abstract) The objective of this chapter is to obtain a classification of Finslerian manifolds. Let (M, g) be a Finslerian manifold of dimension n, and W(M) the fibre bundle of unit tangent vectors to M. The curvature form of the Finslerian connection (Cartan) associated to (M, g) is a two from on W(M) with values in the space of skew-symmetric endomorphisms of the tangent space to M. It is the sum of three two forms of type (2, 0), (1, 1) and (0, 2) whose coefficients R, P and Q constitute the three curvature tensors of the given connection. In the first part we study the Landsberg manifolds, manifolds with minimal fibration and Berwald manifolds.

The manifold M is called a Landsberg manifold if P vanishes everywhere. This condition is equivalent to the vanishing of the covariant derivative in the direction of the canonical section v: M $\to$ V(M) of the torsion tensor. For a Riemannian metric (0,2) on V(M) this condition means that for every $x \in M$ the fibre $p^{-1}(x)$ becomes a totally geodesic manifold where p: V(M) $\to$ M (see [5 and §7]). We examine the case when V(M) $\to$ M is of minimal fibration as well as when M is a Berwald manifold. When M is compact and without boundary we put some global conditions on the first curvature tensor R or flag curvature of the Cartan connection. In the second part we study by deformations the metric of compact Finslerian manifolds in order that their indicatrix become Einstein manifolds.

Let M be a Finslerian manifold. We suppose M to be compact and without boundary. The torsion and curvature tensors of a Finslerian connection satisfy the Bianchi identities, one of which is the following (see [8.17, chap II])

$$(0.1)\qquad \nabla^{\bullet}_{m} R^{i}_{jkl} + T^{r}_{km} R^{i}_{jrl} + T^{r}_{lm} R^{i}_{jkr} + \nabla_{k} P^{i}_{jlm} - \nabla_{l} P^{i}_{jkm} + P^{i}_{jkr} \nabla_{o} T^{r}_{lm} - P^{i}_{jlr} \nabla_{o} T^{r}_{km} + Q^{i}_{jrm} R^{r}_{okl} = 0$$

We can put on the fibre bundle V(M) a Riemannian metric of the form

(0.2) $$\widetilde{g} = g_{ij}\, dx^i\, dx^j + g_{ij}\, \nabla v^i \nabla\, v^j$$

For all $x \in M$, the fibre $p^{-1}(x)$ is a Riemannian submanifold of V(M). We say that (M, g) is a Landsberg manifold if $p^{-1}(x)$ is a *totally geodesic manifold* (see [4]).

In this case P = 0. Equivalently $\nabla_o T = 0$ where T is the torsion tensor. Similarly, we say that V(M) is a minima fibration if $\nabla_o T^* = 0$, (necessary condition where T^* = trace T) [see §7].

1. Landsberg Manifolds.

Let us denote by $\overline{R}$ the symmetric tensor defined by

$$\overline{R}_{ij} = \overline{F}^{-2} R_{iojo}$$

where o denotes the multiplication contracted by v. The fact that R_{iojo} is symmetric is a consequence of the first identity of Bianchi (8.15 chap II). On the other hand let $a_{ij}(z)$ be the symmetric tensor defined by

(1.1) $$a_{ij}(z) = F^2\, \nabla^{\bullet}_{i}\, T_j + v_i\, T_j + v_j\, T_i - F^2\, Q_{ij}$$

where $Q_{ij} = Q^r_{irj}$ and the last is the third curvature tensor of the Cartan connection :

(1.2) $$Q^i_{jkl} = T^i_{ls}\, T^s_{jk} - T^i_{ks}\, T^s_{jl}$$

Lemma 1. *For a compact Finslerian manifold without boundary we have*

(1.3) $$\langle \nabla_o T, \nabla_o T \rangle = \int_{W(M)} (\nabla_o T,\ \nabla_o T)\ \eta = - \langle a, \overline{R} \rangle$$

where $\langle\ \rangle$ and (,) denote respectively the global and local scalar products over W(M).

Proof. Let us multiply the two sides of the Bianchi identity (0.1) by v^l and v^j. We then obtain

$$(1.4)\quad \nabla_o \nabla_o T^i_{km} + T^r_{km} R^i_{oro} + \nabla^{\bullet}_m R^i_{jkl} v^l v^j = 0$$

Whence, on multiplying the two sides by T_i^{km},

$$(1.5)\quad T_i^{km} \nabla_o \nabla_o T^i_{km} + T_i^{km} T^r_{km} R^i_{oro} + T_i^{km}(\nabla^{\bullet}_m R^i_{jkl}) v^j v^l = 0$$

The first term can be written, on taking into account the divergence formula

$$(1.6)\quad T_i^{km} \nabla_o \nabla_o T^i_{km} = \nabla_o (T_i^{km} \nabla_o T^i_{km}) - (\nabla_o T_i^{km} \nabla_o T^i_{km})$$
$$= \frac{1}{2} \nabla_o \nabla_o (T, T) - (\nabla_o T, \nabla_o T)$$
$$= \text{Div on } W(M) - (\nabla_o T, \nabla_o T)$$

Similarly the last term of (1.5) can be written, on taking into account the formula of vertical divergence (7.6 Chapter III)

$$(1.7)\quad T_i^{km}(\nabla^{\bullet}_m R^i_{jkl}) v^j v^l = \nabla^{\bullet}_m [(T_i^{km} R^i_{jkl}) v^j v^l] - (\nabla^{\bullet}_m T_i^{km} R^i_{jkl} v^j v^l)$$
$$- T_i^{km} R^i_{jkl}(\delta^j_m v^l + \delta^l_m v^j)$$
$$= -(\nabla^{\bullet}_i T^k + T_m T_i^{km}) R^i_{oko} + \text{div on } W(M),$$

On substituting (1.7) and (1.6) in (1.5) and taking into account the expression of the Ricci tensor Q_{jl} defined from (1.2), we obtain

$$(1.8)\quad (\nabla_o T, \nabla_o T) = -(\nabla^{\bullet}_i T^k - Q^k_i) R^i_{oko} + \text{div on } W(M)$$

We can add the terms $v^k T_i + v_i T^k$. The expression of the right hand side does not change since

$$v_i R^i_{oko} = 0 = v^k R^i_{oko}$$

(1.9) $(\nabla_o T, \nabla_o T) = -(a, \bar{R}) + \text{Div on } W(M)$

where a is a 2-tensor defined by (1.1). Since we suppose M to be compact, and without boundary, by integrating over W(M), we obtain the lemma. From the above lemma we obtain the following theorem.

We say that a Finslerian manifold has a non-negative curvature in the large sense if the scalar product of the flag curvature by a symmetric tensor of order two ($\bar{R}$, a) is non-negative.

Theorem 1. *Let (M, g) be a compact Finslerian manifold without boundary. If the symmetric tensor $\bar{R}$ is such that $\langle a, \bar{R} \rangle$ is everywhere non-negative, then (M, g) is a Landsberg manifold.*

2. Finslerian Manifolds with minima fibration.

We have seen that the necessary condition for the fibre of V(M) for every $x \in M$ be a minimal submanifold is that $\nabla_o T_* = 0$. (See [5]). In fact $\nabla_o T_*$ is the trace of the fundamental form of the submanifold $p^{-1}(x)$ of V(M) (See [5]).

We are going to study in this paragraph when M is compact the conditions for $\nabla_o T_* = 0$. On contracting i and m in the formula (1.4) we obtain:

(2.1) $$\nabla_o \nabla_o T_k + T^r{}_{ki} R^i_{oro} + v^j v^l \nabla_i R^i_{jkl} = 0$$

From it, we deduce, on multiplying the two sides by T^k and on using the divergence formula

(2.2) $(\nabla_o T_*, \nabla_o T_*) + \text{div on } W(M) = T^k T^r{}_{ki} R^i_{oro} + T^k \nabla_i R^i_{jkl} v^j v^l$

Let us put

(2.3) $$Y_i = F^{-1}\, T^k R_{ioko}$$

(2.4) $$-\dot{\delta} Y = \nabla^{\bullet}_i (T^k R^i_{oko}) + T^i T^k R_{ioko}$$
$$= \nabla^{\bullet}_i T^k R^i_{oko} - T^k R_{ok} + T^i T^k R_{ioko} + v^j v^l T^k \nabla^{\bullet}_i R^i_{jkl}$$

where $R_{ok} = R^i_{oik}$

On putting the last term of (2.4) in (2.2) and on simplifying we have

(2.5)
$$(\nabla_o T_*, \nabla_o T_*) = -(\partial^{\bullet}_i T^k + T_i T^k) R^i_{oko} + T^k R_{ok} + \text{Div on } W(M)$$

First, let us note (see [3])

$$R^i_{okl} = H^i_{okl} = \frac{1}{3}(\partial^{\bullet}_l H^i_{oko} - \partial^{\bullet}_k H^i_{olo})$$

Whence, by contracting i and l:

$$T^k R_{ok} = -T^k R^i_{oki} = -\frac{1}{3} T^k \partial^{\bullet}_i H^i_{oko} + \frac{1}{3} T^k \partial^{\bullet}_k H^i_{oio}$$

$$= -\frac{1}{3} T^k \partial^{\bullet}_i g^{il} H_{loko} - \frac{1}{3} T^k g^{il} \partial^{\bullet}_i H_{loko} + \frac{1}{3} T^k \partial^{\bullet}_k H^i_{oio}$$

$$= \frac{2}{3} T^k T^l H_{loko} - \frac{1}{3} g^{il} \partial^{\bullet}_i (T^k H_{loko}) + \frac{1}{3} T^k \partial^{\bullet}_k H^i_{oio} + \frac{1}{3} g^{il} \partial^{\bullet}_i T^k H_{loko}$$

(2.6) $$T^k R_{ok} = \frac{2}{3} T^k T^l H_{loko} + \frac{1}{3} g^{il} \partial^{\bullet}_i T^k H_{loko}$$
$$+ \frac{1}{3} T^k \partial^{\bullet}_k H^i_{oio} + \text{div on } W(M)$$

Now the last term can be written

(2.7) $$\frac{1}{3} T^k \partial^{\bullet}_k H^i_{oio} = \frac{1}{3} g^{kl} \partial^{\bullet}_k (T_l H^i_{oio}) - \frac{1}{3} g^{kl} \partial^{\bullet}_k T_l H^i_{oio}$$

The first term on the left hand side is a divergence. On taking into account the relations (2.6) and (2.7), the relation (2.5) becomes

(2.8)

$$(\nabla_o T_*, \nabla_o T_*) = -\frac{1}{3}(2\,\partial^{\bullet}_{i} T^k + T_i T^k + h^k_i g^{rs} \partial^{\bullet}_{r} T_s) R^i_{oko} + \text{div on } W(M)$$

where $h^k_i = \delta^k_i - u^k u_i (u = v/F)$. We can add the terms $F^{-1}(u^k T_i + u_i T^k)$ to what is in the parenthesis of the right hand side of (2.8). For as a result of the properties of the tensor R^i_{oko}, the relation does not change. Let us put

$$(2.9)\quad F^{-2} b^k_i = 2[\partial^{\bullet}_{i} T^k + F^{-1}(u^k T_i + u_i T^k)] + T_i T^k + h^k_i g^{rs} \partial^{\bullet}_{r} T_s$$

Thus (2.8) becomes

$$(2.10)\quad (\nabla_o T_*, \nabla_o T_*) = -\frac{1}{3}(b, \overline{R}) + \text{Div on } W(M)$$

Now M is compact, without boundary. By integration on W(M) we obtain

$$(2.11)\qquad \langle \nabla_0 T_*, \nabla_0 T_* \rangle = -\frac{1}{3}\langle b, \overline{R} \rangle$$

Theorem 2. *Let (M, g) be a compact Finslerian manifold without boundary. If the symmetric tensor $\overline{R}$ is such that $\langle b, \overline{R} \rangle$ is everywhere non-negative, then (M, g) is a manifold with minimal fibration($\nabla_o T_* = 0$)*

3. Case of Isotropic Manifolds.

Let (M, g) be a Finslerian manifold with minima fibration ($\nabla_o T_* = 0$) and isotropic, that is to say (see chapter VI)

(3.1) $R^i_{ojo} = \mathrm{KF}^2\, h^i_j,\ (h^i_j = \delta^i_j - u^i u_j),\qquad (\mathrm{K} \neq 0)$

where δ^i_j is the Kronecker symbol and K is *different from zero.* By Bianchi identity (0.1) (see[1]) we obtain

(3.2) $\mathrm{F}\dfrac{\partial K}{\partial v^m} = \dfrac{3}{(n+1)}\mathrm{K}\,\sigma_m$ where $\sigma_m = -\mathrm{A}_m = -\mathrm{FT}_m$

where T is the trace co-vector of torsion. Let $x_0 \in$ M a fixed point and
C: R$\rightarrow S_{x_0}$ a differential map in the indicatrix S_{x_0} such that C(t) is the trajectory of the vector field σ (t) = -A(t). We have

(3.3) $$\frac{dC}{dt} = \sigma\ (\mathrm{t}) = \frac{du}{dt}$$

Now the coordinates in the fiber S_{x_0} are the u^i ($\|u\| = 1$) where $v^i = \mathrm{F}u^i$.
We have

$$\frac{\partial K}{\partial u^i} = \mathrm{F}\,\frac{\partial K}{\partial v^i}$$

Thus, by (3.2) and (3.3), we have for the fiber at x_0

(3.4) $$\frac{dK}{dt} = \frac{\partial K}{\partial u^m}\frac{du^m}{dt} = \mathrm{F}\,\frac{\partial K}{\partial v^m}\frac{du^m}{dt} = \mathrm{F}\,\frac{\partial K}{\partial v^m}\sigma^m$$
$$= \frac{3}{(n+1)}\mathrm{K}(\sigma,\sigma)$$

On supposing K to be different from zero, we deduce from it

$$\frac{d}{dt}\log \mathrm{K} = \frac{3}{(n+1)}(\sigma,\sigma)$$

Hence the solution

$$K(t) = K(0) \exp\left(\frac{3}{(n+1)} \int_0^t (\sigma, \sigma) dt\right)$$

When $t \to \infty$, $K(t) \to \infty$. Since (σ, σ) and $K(t)$ are bounded functions on the compact S_{x_o}. This is impossible since K is constant. Since K is different from zero, we have $\sigma_m = -A_m = 0$.

Therefore (M, g) is a Riemannian manifold.

Theorem 3. *A Finsleran manifold with curvature isotropic and with minimal fibration ($\nabla_o T_* = 0$) is Riemannian. A fortiori, it is so if (M, g) is an isotropic Landsberg manifold with $K \neq 0$.*

Corollary. *A Finslerian surface (dim M = 2) with minimal fibration ($\nabla_o T_* = 0$), $K \neq 0$ is Riemannian.*

4. Calculation of $(\delta A)^2$ when (M, g) is a manifold with minima fibration.

We suppose

(4.1) $$\nabla_0 T_* = 0$$

By vertical derivation we obtain

(4.2) $$\nabla_j T_i = - \nabla_0 \dot{\nabla}_j T_i$$

Now A = FT ; so we have, using (4.1), (4.2).

(4.3) $$-\delta A = F \nabla_i T^i = -F \nabla_0 \dot{\nabla}_j T^j$$

Hence

$$(\delta A)^2 = F^2 \nabla_0 \dot{\nabla}_i T^i \nabla_0 \dot{\nabla}_j T^j$$

$$= F^2 \nabla_0 [\nabla_i T^i \nabla_0 \dot{\nabla}_j T^j] - F^2 \dot{\nabla}_i T^i \nabla_0 \nabla_0 \dot{\nabla}_j T^j$$

$$(4.4) \qquad = \mathrm{Div} - F^2 \dot{\nabla}_i [T^i \nabla_0 \nabla_0 \dot{\nabla}_j T^j] + F^2 T^i [\nabla_i \nabla_0 \dot{\nabla}_j T^j + \nabla_0 \nabla_i \dot{\nabla}_j T^j]$$

$$= \mathrm{Div} + F^2 T^i \nabla_i \nabla_0 \dot{\nabla}_j T^j$$

On using the identity of Ricci (§ 9 Chapter I) we get

$$\nabla_i \nabla_0 (\dot{\nabla}_j T^j) = \nabla_0 \nabla_i \dot{\nabla}_j T^j - \dot{\nabla}_m \nabla_j T^j R^m_{oio}$$

Putting this in (4.4), we get

$$(4.5) \qquad (\delta A)^2 = \mathrm{Div} - F^2 T^i \dot{\nabla}_m \dot{\nabla}_j T^j R^m_{oio}$$

Let us put

$$(4.6) \qquad A = FT^*,\ B_m = F \dot{\partial}_m f,\ f = F^2 \dot{\nabla}_j T^j,\ u = v/F$$

The relation (4.5) becomes

Lemma 2. *For a manifold with minima fibration we have*

$$(4.7) \qquad (\delta A)^2 = -\bar{R}(A, B) + \mathrm{Div} \text{ on } W(M)$$

where δ is the co-differential and A and B are defined by (4.6) and $\bar{R}(A,B) = (R(A, u)u, B)$.

5. Case where (M, g) is a Landsberg Manifold. The Calculation of $\|\nabla_i A_j\|^2$

Let now (M, g) be a Landsberg manifold. Then $\nabla_0 T_* = 0$. Hence the horizontal covariant derivation of A_i is symmetric. On putting $R_{rj} = R^i_{rij}$ and using the Ricci identity we obtain

$$(5.1)\quad \|\nabla_i A_j\|^2 = F^2 \nabla^i T^j \nabla_i T_j = F^2 \nabla^i (T^j \nabla_i T_j) - F^2 T^j \nabla_i \nabla_j T^i$$
$$-F^2 T^j [\nabla_j \nabla_i T^i + T^r R_{rj} - \dot{\nabla}_r T^i R^r_{oij}] + \text{Div on } W(M)$$
$$= -F^2 \nabla_j (T^j \dot{\nabla}_i T^i) + F^2 \nabla_j T^j \dot{\nabla}_i T^i - F^2 R_{rj} T^r T^j$$
$$+ F^2 T^j \dot{\nabla}_r T^i R^r_{oij} + \text{div on } W(M)$$

$$= (\delta A)^2 - F^2 R_{ij} T^i T^j + F^2 \dot{\nabla}^i T_r T^j R^r_{oij} + \text{div on } W(M)$$

Now $\dot{\nabla}_r T_i$ is symmetric. So, the second term becomes

$$(5.2)\quad F^2 \dot{\nabla}^i T_r T^j R^r_{oij} = F \dot{\nabla}^i [T^j T_r R^r_{oij} F] - T^j T_r R^r_{ooj}$$
$$+ F^2 R_{ij} T^i T^j - F^2 T^j T_r v^m \dot{\nabla}^i R^r_{mij}$$

Since the tensor P = 0, by (0.12) the last term of the right hand side vanishes. Hence (5.2) becomes, on taking into account (0.11)

$$(5.3)\quad F^2 g^{ik} \dot{\nabla}_i T_r T^j R^r_{okj} = -(n-2) R_{iojo} T^i T^j + F^2 R_{ij} T^i T^j$$
$$+ \text{Div on } W(M)$$

On putting this relation in (5.1) we obtain :

Lemma 3. *Let (M,g) be a Landsberg manifold of dimension n. Then we have*

$$(5.4)\quad \|\nabla_i A_j\|^2 = (\delta A)^2 - (n-2)(\overline{R}(A, A)) + \text{Div on } W(M)$$

6. Case of Compact Berwald Manifolds

Let (M, g) be a compact Landsberg manifold without boundary. The tensor T satisfies $\nabla_0 T = 0$. By vertical derivation we conclude that the covariant horizontal derivative of T is completely symmetric. We calculate its square:

$$
\begin{aligned}
(6.1)\quad & F^2\, \nabla^l T^{ijk}\ \nabla_l T_{ijk} \\
&= F^2\ \nabla^j T^{ikl}\, \nabla_l T_{ijk} \\
&= F^2 \nabla^j [T^{ikl} \nabla_l T_{ijk}] - F^2 T^{ikl} \nabla_j \nabla_l\, T^{\ j}_{ik} \\
&= \mathrm{Div} - F^2\, T^{\ l}_{ik}\, [\nabla_l \nabla_j T^{jik} + T^{rik}\, R_{rl} + T^{jrk}\, R^{\ i}_{rjl} + T^{jir}\, R^{\ k}_{rjl} \\
&- \nabla^{\bullet}_{r}\, T^{jik}\, R^{\ r}_{ojl}\,]
\end{aligned}
$$

On using the symmetric properties of the tensor T we obtain

$$
\begin{aligned}
(6.2)\ \left\| F\nabla_l T_{ijk} \right\|^2 &= \mathrm{Div} - F^2\, \nabla_l\, [T^{\ l}_{ik}\, \nabla_j\, T^{jik}\,] + F^2\, \nabla_l\, T^{\ l}_{ik}\, \nabla_j\, T^{jik} \\
&- F^2\, T^{\ l}_{ik}\, T^{rik}\, R_{rl} - 2F^2 T^{\ l}_{ik}\, T^{jrk}\, R^{\ i}_{rjl} + F^2 T^{\ l}_{ik}\, \nabla^{\bullet}_{r}\, T^{jik}\, R^{\ r}_{ojl} \\
= \mathrm{Div} + \left\| \nabla_l A_k \right\|^2 &- F^2 T^{\ l}_{ik}\, T^{rik}\, R_{rl} - 2F^2\, T^{\ l}_{ik}\, T^{jik}\, R^{\ i}_{rjl} + F^2\, T^{\ l}_{ik}\, \nabla^{\bullet j}\, T^{ik}_{\ \ r}\, R^{\ r}_{ojl}
\end{aligned}
$$

where we have used the fact the vertical covariant derivative of the tensor T is also completely symmetric. Now the last term becomes

$$
\begin{aligned}
(6.3)\quad F^2 T^{\ l}_{ik}\, \nabla^{\bullet j} T^{ik}_{\ \ r}\, R^{\ r}_{ojl} &= F\, \nabla^{\bullet j} [F\, T^{\ l}_{ik}\ T^{ik}_{\ \ r}\ R^{\ r}_{ojl}\,] \\
&- T^{\ l}_{ik}\ T^{ik}_{\ \ r}\ R^{\ r}_{ool} - F^2 T^{\ l}_{ik}\ T^{ik}_{\ \ r}\ \nabla^{\bullet j}\, R^{\ r}_{ojl}
\end{aligned}
$$

On taking into account the formulas of vertical divergence (0.10) let us put

$$
Y_j = F\ T^{\ l}_{ik}\ \ T^{ik}_{\ \ r}\ R^{\ r}_{ojl}
$$

$$
\hat{Y}_j = Y_j - u_j\, Y_0/F, \qquad\qquad (u_j = v_j/F)\ Y_0 = v^i Y_i
$$

We have $\quad -\dot{\delta}\,\hat{Y} = \mathrm{F}\ \nabla^{\bullet j}\ \mathrm{Y}_j - (\mathrm{n}\text{-}1)\ \dfrac{Y_0}{F} + \mathrm{F}\mathrm{T}^{j}\mathrm{Y}_j$

Thus the relation (6.3) becomes

$$(6.4)\quad \mathrm{F}^2 T^l_{ik}\ \nabla^{\bullet j}\ T^{ik}_r\ R^r_{ojl} = -\dot{\delta}\,\hat{Y} - \mathrm{F}^2\mathrm{T}^j T^l_{ik}\ T^{ik}_r\ R^r_{ojl} + (\mathrm{n}\text{-}2)\ T^l_{ik}\ T^{ik}_r\ R^r_{ool}$$
$$- \mathrm{F}^2 T^l_{ik}\ T^{ik}_r\ \nabla^{\bullet j}\ R^r_{ojl}$$

Now, since P = 0, from Bianchi identity (0.1) we obtain

$$\mathrm{v}^j \nabla^{\bullet k}\ R^i_{jkl} + \mathrm{T}^r\ R^i_{orl} = 0$$

Thus the last term of (6.4) becomes

$$\nabla^{\bullet j}\ R^r_{ojl} = -\,R^r_l - \mathrm{T}^s\ R^r_{osl}$$

Therefore

$$\mathrm{F}^2\ T^l_{ik}\ \ \nabla^{\bullet j}\ T^{ik}_r\ R^r_{ojl} = -\dot{\delta}\,\hat{Y} + (\mathrm{n}\text{-}2)\ T^l_{ik}\ T^{ik}_r\ R^r_{ool} + \mathrm{F}^2\ T^l_{ik}\ T^{ikr}\ R_{rl}$$

Let us put this expression in (6.2)

$$(6.5)\ \left\| F\nabla_l T_{ijk} \right\|^2 = \left\| \nabla_l A_k \right\|^2 + (\mathrm{n}\text{-}2)\ T^l_{is}\ T^{is}_r\ R^r_{ool}$$
$$- 2\mathrm{F}^2 T^l_{is}\ T^{jrs}\ R^i_{rjl} + \text{div on W(M)}$$

On taking into account the expression of the curvature tensor Q defined by (1.2) and the properties of the tensors R and T , the last term of the right hand side of (6.5) becomes

$$-2\ \mathrm{F}^2\ T^l_{is}\ T^{jrs}\ R^i_{rjl} = -\ \mathrm{F}^2 Q^i_{jkl}\ R^{jkl}_i$$

Now P= 0. So we have (see [4] p.289)

$$R^i_{jkl} = H^i_{jkl} + T^i_{jr}\ \ R^r_{okl}$$

Finally (6.5) becomes

(6.6)

$$\|F\nabla_l T_{ijk}\|^2 = \|\nabla_l A_k\|^2 \text{ -(n-2) } T^l_{is}\, T^{is}_r\, R^r_{olo} - F^2\, Q^i_{jkl}\, H_i^{jkl} + \text{div on W(M)}$$

Let us note that the tensor H is (see [4])

$$H^i_{jkl} = \frac{1}{3}\, \partial^{\bullet}_j\, (\partial^{\bullet}_l\, H^i_{oko} - \partial^{\bullet}_k\, H^i_{olo})$$

So the last term of (6.6) becomes :

$$\text{F}^2\, Q^i_{jkl}\, H_i^{jkl} = \frac{2}{3}\text{F}^2\, Q_i^{jkl}\, \partial^{\bullet}_j\, \delta^{\bullet}_l\, H^i_{oko}$$

$$(6.7)\quad = \frac{2}{3}\text{Fg}^{\text{jm}}\, \partial^{\bullet}_j\, [\text{F}\, Q_{im}{}^{kl}\, \delta^{\bullet}_l\, H^i_{oko}] - \frac{2}{3}\text{Fg}^{\text{jm}}\, \partial^{\bullet}_j\, [\text{F}\, Q_{im}{}^{kl}]\, \delta^{\bullet}_l\, H^i_{oko}$$

$$= \text{Div - } \Psi_i^{lk}\, \delta^{\bullet}_l\, H^i_{oko}$$

We have put, on using the fact that $Q_{im}{}^{kl}$ is skew-symmetric with respect to (k,l)) and (i, m)

$$(6.8)\qquad \Psi_i^{lk} = \frac{2}{3}\, \text{g}^{\text{jm}}\, \partial^{\bullet}_j (\text{F}^2\, Q_{mi}{}^{lk})$$

Since Ψ_i^{lk} is homogeneous of degree (-1), (6.7) becomes

$$(6.9\ \text{F}^2 Q^i_{jkl}\, H_i^{jkl} = \text{Div-Fg}^{\text{ls}}\, \delta^{\bullet}_l (\Psi^k_{is}\, H^i_{oko}/\text{F}) + \frac{2}{3}\, Q^k_i\, H^i_{oko}$$

$$+ \text{g}^{\text{ls}}\, \delta^{\bullet}_l (\Psi^k_{is})\, H^i_{oko}$$

$$= \text{Div} + [\frac{2}{3}\text{n}\, Q^k_i + \text{g}^{\text{ls}}\, \delta^{\bullet}_l\, \Psi^k_{is}]\, H^i_{oko}$$

Let us put

$$(6.10)\qquad C^k_i = \frac{2}{3}\text{n}\, Q^k_i + \text{g}^{\text{ls}}\, \delta^{\bullet}_l\, \Psi^k_{is}$$

and

(6.11) $$A_i^k = F^2\,(C_i^k + (n-2)\,T_{js}^k\,T_i^{js})$$

Thus we can the relation (6.6)

(6.12) $$\|F\nabla_l T_{ijk}\|^2 = \|\nabla_l A_k\|^2 - A_i^k\,H_{oko}^i + \text{div on } W(M)$$

On using the lemmas 2 and 3 and on putting

$$J_i^k = A_i^k + (n-2)\,A_i A^k + A^k B_i$$

where B_i is defined by (4.3), we finally obtain

$$\|F\nabla_l T_{ijk}\|^2 = -\,F^{-2}\,J_i^k\,H_{oko}^i + \text{div on } W(M)$$

On integrating on W(M) we have

Theorem 4. *Let (M, g) be a compact Landsberg manifold without boundary, dim M = n > 2. If* $J_i^k\,H_{oko}^i$ *is non-negative everywhere on W(M) then (M, g) is a Berwald manifold.*

7. Finslerian manifolds whose fibres are totally geodesic or minima

The Finslerian connection defines at each point $z \in V(M)$ a decomposition of the tangent space to V(M) at this point. We have $T_zV(M) = H_z \oplus V_z$ where H_z is the horizontal and V_z the vertical space. At the point $z \in V(M)$, the Pfaffian derivatives ($\partial_k = \delta_k - \Gamma^{*r}_{0k}\,\dot{\delta}_r$, $\dot{\delta}_k$) define a frame adapted to the decomposition of $T_zV(M)$. Let us put the Riemannian metric on V(M):

(7.1) $$ds^2 = g_{ij}\,dx^i\,dx^j + g_{ij}(z)\,\nabla v^i \nabla v^j\,,\ z \in V(M)$$

Let EV(M) be the principal fibre bundle of linear frames on V(M) with the structure group GL(2n, **R**). Let $\hat{D}$ be the Riemannian connection associated to (7.1). This connection has no torsion and $\hat{D}G = 0$. Let π^{α}_{β} (α, β=i, $\bar{i}$=1,2, ...n) be the matrix of this connection relative to the adapted frame, we have

$$(7.2) \qquad \hat{D}(\partial\beta) = \pi^{\alpha}_{\beta}\, \partial_{\alpha}, \qquad (\pi^{\alpha}_{\beta} = \Gamma^{\alpha}_{\beta\lambda}\, \sigma^{\lambda})$$

where $\sigma^{\lambda} = (dx^i, \nabla v^j)$.

$\hat{D}$ being Riemannian, we have

$$(7.3) \quad \Gamma^{\gamma}_{\alpha\beta} = \frac{1}{2} G^{\gamma\lambda}(\partial_{\alpha} G_{\beta\lambda} + \partial_{\beta} G_{\lambda\alpha} - \partial_{\lambda} G_{\alpha\beta}) - \frac{1}{2}\{G^{\gamma\lambda}[\partial_{\beta}, \partial_{\lambda}]_{\alpha} + [\partial_{\alpha}, \partial_{\beta}]^{\gamma} - G^{\gamma\lambda}[\partial_{\lambda}, \partial_{\alpha}]_{\beta}\}$$

where the bracket $[\partial_{\alpha}, \partial_{\beta}]$ is defined by

$$(7.4) \qquad [\partial_{I}, \partial_{j}] = -R^{r}{}_{oij}\, \delta_{\dot{r}}$$

$$(7.5) \qquad [\partial_{I}, \delta_{\dot{i}}] = G^{r}{}_{ij}\, \delta_{\dot{r}}$$

$$(7.6) \qquad [\delta_{\dot{i}}, \delta_{\dot{j}}] = 0$$

where the $G^{r}{}_{ij}$ are the coefficients of the Berwald connection associated to g. Calculating the right hand side of (7.3) and taking note of the bracket expression we obtain the 1-form of the Riemannian connection π^{α}_{β} with respect to the adapted frames :

$$\pi^{i}_{j} = \omega^{i}_{j} + \frac{1}{2} g^{ih} R_{okhj} \nabla v^{k}$$

$$(7.7) \qquad \pi^{\bar{i}}_{j} = -(T^{i}{}_{jk} - \frac{1}{2} R^{i}{}_{ojk})dx^{k} - \nabla_0 T^{i}{}_{jk} \nabla v^{k}$$

$$\pi^{i}_{\bar{j}} = (T^{i}{}_{jk} + \frac{1}{2} R_{oj}{}^{i}{}_{k})dx^{k} + \nabla_0 T^{i}{}_{jk} \nabla v^{k} \quad \pi^{\bar{i}}_{\bar{j}} = \omega^{i}_{j}$$

where ω^{i}_{j} represents the 1-form of the Finslerian connection associated to g_{ij} and where T and R are the torsion and curvature tensors of ∇. Let $x = x_0$ be a fixed point of M and the fibre manifold $p^{-1}(x_0) = V^n$ a submanifold of V(M) with the induced Riemannian metric :

(7.8) $$d\sigma^2 \,|_{\,p^{-1}(x_0)} = g_{ij}(x_0, v)\, dv^i dv^j$$

Let $\dot{X}$ and $\dot{Y}$ be two tangent vectors in $(x_0, v) \in V^n$ to $p^{-1}(x_0)$ we have

(7.9) $$\hat{D}_{\dot{Y}}\, \dot{X} = \dot{D}_{\dot{Y}}\, \dot{X} + A(\dot{Y}, \dot{X}),$$

where $\dot{D}$ is the induced connection and A the second fundamental form of the submanifold $p^{-1}(x_0)$. Let us make explicit the right hand side. If $\dot{X} = (\delta_{\dot{j}})$ and $\dot{Y} = (\delta_{\dot{k}})$ we have, with respect to the adapted frame

$$\hat{D}_{\delta_{\dot{k}}}\, \delta_{\dot{j}} = \pi^{\alpha}_{\dot{j}}(\delta_{\dot{k}}) = \pi^{\bar{i}}_{\bar{j}\bar{k}}\, \delta_{\dot{i}} + \pi^{i}_{\bar{j}\bar{k}}\, \delta_i$$

Taking into account (7.7) we obtain

(7.10) $$\hat{D}_{\delta_{\dot{k}}}\, \delta_{\dot{j}} = T^i{}_{jk}\,(x_0, v)\, \delta_{\dot{i}} + \nabla_0\, T^i{}_{jk}\,(x_0, v)\partial_i$$

where the vectors ∂_i and $\delta_{\dot{i}}$ are orthogonal with respect to the metric (7.1) $G(\delta_{\dot{i}}, \partial_i) = 0$. From this formula it follows immediately that for $V^n = p^{-1}(x_0)$ to be a totally geodesic (respectively minima) submanifold, it is necessary and sufficient that $\nabla_0\, T^i{}_{jk} = 0$ (respectively necessary $g^{jk}\nabla_0\, T^i{}_{jk} = \nabla_0\, T^i = 0$). This condition is equivalent to the vanishing of the second curvature tensor P of the Finslerian connection.

Theorem. *In order that the fibres of $p: V(M) \to M$ be totally geodesic (respectively minima) it is necessary and sufficient (respectively necessary) that the second curvature tensor of the Finslerian connection P (respectively $\nabla_0\, T_i = 0$) is zero [5].*

II COMPACT FINSLERIAN MANIFOLDS WHOSE INDICATRIX IS AN EINSTEIN MANIFOLD

1. The first variational of $\dot{I}(g_t) = \int_W F^2 Q_t \eta$

Let $F(g_t)$ be a 1- parameter family of Finslerian metric We denote by $F^0(g_t)$ the sub-family of metrics such that for every $t \in [-\varepsilon, \varepsilon]$ the volume of the unitary tangent fibre bundle of M corresponding to g_t is equal to 1. Let us denote by Q_t the scalar curvature corresponding to the third curvature tensor Q (1.2.I) of the Finslerian connection. Let us look for $g_t \in F^0(g_t)$ such that the integral $I(g_t)$ defined by

(1.1) $$I(g_t) = \int_W F^2 Q_t \eta$$

(1.2) $$\int_W \eta_t = 1 \qquad (Q = g^{ij} Q_{ij})$$

be an extremum with the volume constant equal to 1. *We suppose that the torsion tensor is invariant by deformation.* On taking into account (1.1 II chap IV) we have

(1.3) $$(F^2 Q_t \eta)' = - [F^2 Q_{ij} - F^2 Q g_{ij}) t^{ij} + ((\frac{n}{2} - 1)F^2 Qt(u, u)] \eta$$

since T is invariant by deformation its trace is also invariant by deformation.

Let us put

(1.4) $$n \Psi = (\frac{n}{2} - 1)F^2 Q$$

and use the formula (4.3 II chap IV) So (1.3) becomes

(1.5) $$(F^2 Q_t \eta)' = - q_{ij} t^{ij} \eta$$

with

(1.6) $$q_{ij} = F^2 Q_{ij} + (\Psi - F^2 Q)g_{ij} + \frac{1}{n-2} F^2 \partial^{\bullet\bullet}_{ij} \Psi$$

From (1.4) we have

(1.7) $$(\Psi - F^2 Q) = -(\frac{n+2}{2n})F^2 Q$$

Since we have assumed the volume of W(M) to be constant we have the relation (3.9, II, chap IV).
If for $t = 0$; q_{ij} satisfies

(1.8) $$q_{ij} = F^2 Q_{ij} - \frac{n+2}{2n} F^2 Q g_{ij} + \frac{1}{n-2} F^2 \partial^{\bullet\bullet}_{ij} \Psi = c\, g_{ij} + b u_i u_j$$

where c and b are arbitrary constants. Then the first derivative of I(g) at this point $g_0 \in F^0(g_t)$ vanishes and $g_0 \in F^0(g_t)$ for $t = 0$ is a critical point of I(g). On multiplying the two sides of (1.8) by u^i and u^j on one side and by g^{ij} on the other part we obtain successively on taking account of the property of the tensor Q ($Q_{oj} = Q_{io} = 0$)

(1.9) $$-\frac{n+2}{2n} F^2 Q = c + b$$

(1.10) $$-\frac{n}{2} F^2 Q + \frac{1}{n-2} F^2 \partial^{\bullet\bullet}_{ij} \Psi = nc + b$$

From the relation (1.9) it follows that at the critical point $F^2 Q$ is constant. On putting it in (1.10) we obtain

(1.11) $$\frac{1}{n-2} F^2 g^{ij} \partial^{\bullet\bullet}_{ij} \Psi = \frac{2nc-(n+1)(n-2)b}{(n+2)}$$

The first part is a divergence (See 7.6 chap III), M being compact and without boundary and volume W(M) = 1 on integrating (1.11) on W(M) we get

(1.12) $$c = \frac{(n-2)(n+1)}{2n} b$$

and from (1.11) we have for $n \neq 2$,

$$F^2 g^{ij} \dot{\partial}_{ij} \Psi = 0$$

On multiplying the two sides of this relation by Ψ, by integration we obtain

(1.13) $$\frac{1}{2} \int_W F^2 g^{ij} \dot{\partial}_i \dot{\partial}_j \Psi^2 \eta = \int_W F^2 g^{ij} \dot{\partial}_i \Psi \dot{\partial}_j \Psi \eta = 0$$

Let us put the expression defined by (1.12) in (1.9). We then have

(1.14) $$F^2 Q = -(n-1)b$$

And by (1.8) we finally get:

(1.15) $$F^2 Q_{ij} = \frac{1}{n-1} F^2 Q (g_{ij} - u_{ij}) = \frac{1}{n-1} F^2 Q\ h_{ij}$$

where $F^2 Q$ is constant. If the indicatrix S_x is represented by $v^i(t^\alpha) = v^i$ ($\alpha = 1, \ldots n-1$), then the metric $\breve{g}$ defined by (0.2) induces on S_x the metric

$$g_{\alpha\beta} = g_{ij} \frac{\partial v^i}{\partial t^\alpha} \frac{\partial v^j}{\partial t^\beta} \qquad \alpha, \beta = 1, \ldots n-1$$

and the curvature tensor of S_x (see [4]) is represented by

$$\dot{R}_{\alpha\beta\gamma\lambda} = \dot{Q}_{\alpha\beta\gamma\lambda} + g_{\alpha\gamma} g_{\lambda\beta} - g_{\alpha\lambda} g_{\beta\gamma}$$

where $\dot{Q}$ is the projection of the curvature tensor Q on the indicatrix S_x

On multiplying by $g^{\alpha\gamma}$:

$$\dot{R}_{\beta\lambda} = F^2 Q_{\beta\lambda} + (n-2)\, g_{\beta\lambda}$$

This is the Ricci tensor of the induced connection (See [4]). But after (1.15) we have by projection on S_x: $F^2 Q_{\beta\lambda} = \dfrac{1}{n-1} F^2 Q g_{\beta\lambda}$

Thus

$$\dot{R}_{\beta\lambda} = [\frac{1}{n-1} F^2 Q + (n-2)] g_{\beta\lambda} \tag{1.16}$$

where $F^2 Q$ is constant. Thus S_x is an Einstein manifold ($n \neq 2$)

Theorem. *Let (M, g) be a deformation of a compact Finslerian manifold without boundary and ($n \neq 2$). At the critical point $g_0 \in F^0$ (g_t) for $t = 0$ of the integral $I(g_t)$, the indicatrix S_x is an Einstein manifold by (1.16) where $F^2 Q$ is constant.*

2. The Second Variational

The derivative under the integral (1.1) is defined by (1.3). We always suppose that the torsion tensor is invariant by deformation. By (1.5) the first derivative is

$$I'(g_t) = \int_{W(M)} A(g)\, \eta \tag{2.1}$$

where

$$A(g) = [F^2 Q_{ij} - F^2 Q g_{ij}) t^{ij} + (\frac{n}{2} - 1) F^2 Q\, t(u, u)] \tag{2.2}$$

In the second derivative appear the expressions g^{ij}, t_{ij}, $t'(u, u)$. We obtain them by deriving the relations (3.9 chap IV) :

$$\int_{W(M)} t'(u, u)\, \eta = - \int_{W(M)} [t(u, u) \operatorname{trace} t - (\frac{n}{2}+1)\, t^2 (u, u)]\, \eta \tag{2.3}$$

$$\int_{W(M)} g^{ij} t'_{ij}\, \eta = \int_{W(M)} [(t, t) - \operatorname{trace} t\, (\operatorname{trace} t - (\frac{n}{2})\, t(u, u)]\, \eta \tag{2.4}$$

On deriving (2.1) and using (2.3) and (2.4) as well as the values at the point t = of F^2 Q $_{ij}$ and F^2 Q defined by (1.15) and (1.14), we obtain after long calculations

$$(2.5)\quad I''(g_o) = -b \int_{W(M)} [(t,t) - (\text{trace } t)^2 - n(\frac{n}{2}-3)\, t^2 (u,u) + (3\frac{n}{2}-1)\text{trace } t\ t(u,u)]\,\eta$$

By an infinitesimal conformal deformation we have

$$(2.6)\qquad t_{ij} = g'_{ij} = 2\,\varphi(x)\, g_{ij}$$

$$(2.7)\qquad I''(g_o) = -\,4n(n-3)b \int_{W(M)} \varphi^2 \eta$$

Theorem *The second variational of the integral $I(g_t)$ where $g_0 \in F^0$ (g_t) is a critical point and defined by the formula (2.5) and for an infinitesimal conformal deformation we have (2.7) at this point point, for n> 3, according as constant scalar curvature is positive, zero or negative.*

In case $n = 3$, $I''(g_o) = 0$, the Weil curvature tensor of the submanifold $p^{-1}(x) \subset V(M)$ vanishes. Since $p^{-1}(x)$ is an Einstein manifold for g_o, it has a constant sectional curvature, and by projection we conclude that the indicatrix has a constant sectional curvature.

CHAPTER VI

Finslerian Manifolds
of
Constant Sectional Curvature [4]

(**abstract**) This chapter is a study of isotropic and constant sectional curvature Finslerian manifolds. We first recall briefly the basics of Finslerian manifolds, define the isotropic manifolds and single out the properties of their curvature tensors. We then give a characterization of Finslerian manifolds with constant sectional curvature, generalizing Schur's classical theorem.
We next determine the necessary and sufficient conditions for an isotropic Finslerian manifold to be of constant sectional curvature. Our conditions bear on the Ricci directional curvature or on the second scalar curvature of Berwald. We show that the existence of normal geodesic coordinates of class C^2 on isotropic manifolds forces them to be Riemannian or locally Minkowskian. We also deal with the case of compact isotropic Finslerian manifolds with strictly negative curvatures.
In chapter III we give a classification of complete Finslerian manifolds with constant sectional curvatures. We prove that all geodesically complete Finslerian manifolds of dimension $n > 2$ with negative constant sectional curvature ($K < 0$) and with bounded torsion vector are Riemannian. We show that all simply connected Finslerian manifolds of dimension $n > 2$ with strictly positive constant sectional curvature and whose indicatrix is symmetric and has a scalar curvature independent *of the direction is* homeomorphic to an n-sphere. In the case when the Berwald curvature H vanishes and torsion tensor as well as its covariant vertical derivative are bounded we prove that the manifold in question is Minkowskian.
In the last chapter we establish the 'axioms of the plane'. By defining the totally geodesic, semi-parallel and auto-parallel Finslerian submanifolds we establish the criteria that permit to identify if a Finslerian manifold is of constant sectional curvature in the Berwald connection (axiom 1), in the Finslerian connection (axiom 2) or is Riemannian (axiom 3).

I Isotropic Finslerian Manifolds

Notation and Recalls

1. Finslerian Manifolds.

Let (x^i) $(i = 1, 2, \ldots n)$ be a local chart of the domain $U \subset M$, and (x^i, v^i) the induced local chart on $p^{-1}(U)$ where $v = v^i \frac{\delta}{\delta x^i} \in T_{pz}(M)$.

Definition. A Finslerian manifold is defined by the data of a function F on TM, satisfying the following conditions

1) $F > 0$ and C^∞ on V(M)
2) $F(x, \lambda v) = \lambda F(x, v)$, $\lambda \in R^+$
3) $g_{ij}(x, v) = \frac{1}{2} \frac{\partial^2 F^2}{\partial v^i \partial v^j}$ is positive definite.

We say the manifold is pseudo-Finslerian if g_{ij} defines non-degenerate quadratic form. Let us recall that g_{ij} is a tensor, homogeneous of degree zero in v. So we have

$$g_{ij}(x, v)\, v^i v^j = F^2$$

where g(,) denotes the local scalar product at $z \in V(M)$. If $g_{ij}(x,v)$ is independent of v, we then have the structure of a Riemannian manifold.

It is shown in [Chap II] that there exists a unique regular connection attached to F such that

1) $\nabla_{\hat{Z}}\, g = 0$

(1.1) 2) $S(X, Y) = 0$

3) $g(T(\dot{X}, Y), Z) = g(T(\dot{X}, Z), Y)$

where $\dot{X} = \mu(V\hat{X})$, $Y = \rho(\hat{Y})$, $Z = \rho(\hat{Z})$, $\hat{X}, \hat{Y}, \hat{Z} \in TV(M)$

The connection ∇ thus defined is called the Finslerian connection. From the preceding conditions it follows that the covariant derivation ∇ is determined by

$$(1.2)\qquad \begin{aligned} 2g(\nabla_{\hat{X}} Y, Z) = {} & \hat{X} g(Y, Z) + \hat{Y} g(X, Z) - \hat{Z} g(X, Y) \\ & + g(\tau(\hat{X}, \hat{Y}), Z) + g(\tau(\hat{Z}, \hat{X}), Y) + g(\tau(\hat{Z}, \hat{Y}), X) \\ & + g(\rho[\hat{X}, \hat{Y}], Z) + g(\rho[\hat{Z}, \hat{X}], Y) + g(\rho[\hat{Z}, \hat{Y}], X) \end{aligned}$$

With respect to the natural frame induced at $z \in p^{-1}(U)$, $\delta_i = \dfrac{\delta}{\delta x_i}$, $\delta_{\dot{i}} = \dfrac{\delta}{\delta v^i}$, we put

$$(1.3)\qquad \nabla_{\delta_j} \delta_i = \Gamma^k_{ij}\, \delta_k \qquad \nabla_{\delta_{\dot{j}}} \delta_i = C^k_{ij}\, \delta_k$$

From the first structure equation, it follows immediately that the torsion tensor T in these co-ordinates coincides with C. From (1.2) we obtain:

$$(1.4)\qquad T_{ijk} = \frac{1}{2}\, \delta_{\dot{k}}\, g_{ij}, \quad T_{ijk} = g_{ir}\, T^r{}_{jk}$$

Taking into account the homogeneity of the right hand side we get

$$(1.5)\qquad T_{ojk} = T_{iok} = T_{ijo} = 0$$

where o denotes the multiplication contracted by v.

Let $\partial_j = \delta_j - \Gamma^{k}_{oj}\,\dot{\delta}_k$ the horizontal vector field over δ_j:
$(\rho(\partial_j) = \delta_j,\ \nabla_{\partial_j} v = 0)$.

Let us put

$$(1.6) \qquad \nabla_{\partial_j} \delta_i . = \overset{*}{\Gamma}{}^{k}_{ij}\,\delta_k$$

By (1.3) we have $\qquad \overset{*}{\Gamma}{}^{k}_{ij} = \Gamma^{k}_{ij} - T^{k}_{ir}\,\Gamma^{r}_{oj}$

Now, by $(1.1)_2$, $\overset{*}{\Gamma}$ is symmetric with respect to the indices i and j. On taking three horizontal vector fields ∂_i, ∂_j, and ∂_k over δ_i, δ_j, and δ_k respectively and on using (1.2) we obtain :
(1.7)

$$2\,\overset{*}{\Gamma}_{kij} = (\partial_i g_{jk} + \partial_j g_{ik} - \partial_k g_{ij}), \quad (\overset{*}{\Gamma}_{kij} = g_{kr}\overset{*}{\Gamma}{}^{r}_{ij} \text{ and } \partial_j = \delta_j - \Gamma^{k}_{oj}\,\dot{\delta}_k)$$

In the same way one shows that exists a connection regular, non-metric and torsionless, attached to F, called the Berwald connection [chap II]. If D is the covariant derivation in this connection there is a relationship between D and ∇:

$$(1.8) \qquad D_{H\hat{X}} Y = \nabla_{H\hat{X}} Y + (\nabla_{\hat{v}} T)(X, Y)$$

$$(1.9) \qquad D_{V\hat{X}} Y = V\hat{X}\,.Y$$

where $\hat{v}$ is horizontal above v ($\rho\,\hat{v} = v$). From (1.8) it follows that D and ∇ define the same splitting on the tangent fibre bundle to V(M). We have

$$(1.10). \qquad \nabla_{\hat{X}} v = D_{\hat{X}} v$$

for all vector fields $\hat{X}$ on V(M). If $\hat{Y}$ is the projectable by p*, then $D_{V\hat{X}} Y = 0$. D admits two curvature tensors denoted by H and P = - G. Let us put

(1.11). $D_{\delta_j}(\delta_i) = G^k_{ij}\delta_k, \quad D_{\dot{\delta}_j}(\delta)_i = 0.$

Then, by (1.8) we have the relation between G and $\overset{*}{\Gamma}$

(1.12). $G^i_{jk}(x, v) = \overset{*}{\Gamma}{}^i_{jk} + \nabla_0 T^i_{jk}$

2. Indicatrices.

For $x_0 \in M$ fixed, the fiber $p^{-1}(x_0) = V_n$ is a differentiable submanifold of V(M). It will be equipped with a Riemannian metric g(X,Y) where X and Y are two vertical tangent vectors to V at $z \in p^{-1}(x_0)$. we call the indicatrix at $x_0 \in M$ the hypersurface S of V_n of the equation

(2.1). $S : F(x_0, v) = 1$

If f is a homogeneous function of degree zero in v, we denote by $\bar{\delta}_i f$ the partial derivatives of f with respect to the unitary vector and we have

(2.2). $\|v\| \frac{\delta f}{\delta v^i} = \bar{\delta}_i f$

where $\|v\|$ is the norm of v. The hypersurface S can be represented by

(2.3). $v^i = v^i(t^\alpha), \quad \alpha = 1, \ldots n-1$

Let f be a function defined on S; its differential is written

$$df(v(t)) = \bar{\delta}_i \cdot f\, dv^i = \partial_\alpha f dt^\alpha = v^i_\alpha \bar{\delta}_i \cdot f\, dt^\alpha$$

whence

(2.4). $\partial_\alpha = v^i{}_\alpha \bar{\delta}_i^\bullet$, $\quad (v^i{}_\alpha = \frac{\partial v^i}{\partial t^\alpha})$

The ∂_α define (n-1) tangent vectors to S, the matrix $(v^i{}_\alpha)$ is of rank (n-1), the induced tensor metric on S becomes

(2.5). $g_{\alpha\beta} = g_{ij}\, v^i{}_\alpha\, v^j{}_\beta$

Let $\dot{v} = v^j\, \partial\dot{}_j$ the tangent vector to the fibre passing through z, on differentiating $F^2(x_0, v) = 1$, we have

$$g_{ij}\, v^i{}_\alpha\, v^j = g(\partial_\alpha, \dot{v}) = 0$$

Thus the vector $\dot{v}$ is normal to (n-1), $v^i{}_\alpha$ tangent vectors to S. In this way, the set $(v^i{}_\alpha, \dot{v})$ define at $z(x_0, v)$ n-vectors linearly independent. Therefore (2.5) can also be written

(2.6). $g_{\alpha\beta} = h_{ij}\, v^i{}_\alpha\, v^j{}_\beta$ $\quad (h_{ij} = g_{ij} - v_i v_j)$

where $v_i = g_{ij} v^j$, $\|v\| = 1$. It then follows

(2.7). $g^{ij} = g^{\alpha\beta}\, v^i{}_\alpha\, v^j{}_\beta + v^i v^j$

Similarly, the matrix $(v^i{}_\alpha, v^i = v^i{}_n)$ is of rank n. On denoting by the same letter its inverse, we have

(2.8). $v^i{}_j\, v^j{}_k = v^i{}_\alpha\, v^\alpha{}_k + v^i{}_n\, v^n{}_k = = \delta^i{}_k \; (v^n{}_i = v_i)$

Thus from (2.6) we get

(2.9). $h_{ij} = g_{\alpha\beta}\, v^\alpha{}_i\, v^\beta{}_j$

At a point $z = (x_0, v) \in V_n = p^{-1}(x_0)$, the coefficients of the Riemannian connection associated to the vertical metric g(X, Y) are $T^i{}_{jk}(x_0, v)$ where T is the torsion tensor of the Finslerian connection. The corresponding Riemannian covariant derivation is determined by

(2.10) $\dot{D}_{\partial_k}\,\partial_j = T^i{}_{jk}\,(x_0, v)\,\partial_i \qquad (\partial_i = \frac{\partial}{\partial v^i})$

Let us denote by $\dot{\nabla}$ the connection induced on S and by a, the second fundamental form of the hypersurface S. If X and Y are two tangent vectors to S, the Gauss equation becomes

(2.11). $\dot{D}_Y X = \dot{\nabla}_Y X + a(X, Y)$

Now $\dot{v}$ is normal to S. So we have

(2.12). $g(\dot{D}_Y X, \dot{v}) = g(a(X, Y), \dot{v}) = -\,g(X, \dot{D}_Y\, \dot{v})$

Now, by (2.10) we have

(2.13). $\dot{D}_Y \dot{v} = Y$

Thus

$$a(X, Y) = -\,g(X, Y)\,\dot{v}$$

The equation (2.11) becomes

(2.14). $\dot{D}_Y X = \dot{\nabla}_Y X - g(X, Y)\,\dot{v}$

If we put in (2.14), $X = \partial_\alpha$, $Y = \partial_\beta$ and $\dot{\nabla}_{\partial_\beta}\,\partial_\alpha = \dot{\Gamma}^\lambda{}_{\alpha\beta}\,\partial_\lambda$ where $\dot{\Gamma}^\lambda{}_{\alpha\beta}$ are the coefficients of the Riemannian connection induced on S, we obtain

(2.15). $\dot{\nabla}_\beta v^i{}_\alpha = -A^i{}_{jk} v^j{}_\alpha v^k{}_\beta - g_{\alpha\beta} v^i, \ (A^i{}_{jk} = FT^i{}_{jk})$

Let us denote by $\dot{R}$ the curvature tensor of the connection induced on S and by $\dot{Q}$ the projection on S of the third curvature tensor of the Finslerian connection. From the relation (2.15) we have

(2.16). $$\dot{R}_{\alpha\beta\gamma\lambda} = F^2\, \dot{Q}_{\alpha\beta\gamma\lambda} + g_{\alpha\gamma}\cdot g_{\lambda\beta} - g_{\alpha\lambda}\cdot g_{\beta\gamma}$$

By contraction we obtain the relation between the scalar curvatures

(2.17). $$\dot{R} = F^2\dot{Q} + (n-1)(n-2)$$

where $\dot{Q} = g^{\alpha\gamma}\, g^{\beta\lambda}\, \dot{Q}_{\alpha\beta\gamma\lambda}$ and $\dot{R} = g^{\alpha\gamma}\, g^{\beta\lambda}\, \dot{R}_{\alpha\beta\gamma\lambda}$

3. Isotropic manifolds

Let $P(v, X) \subset T(M)$ be a two plane generated by two vectors v and $X \in T_{pz}$ where v is the canonical section.

After Berwald the *sectional curvature following* P in the Berwald connection is defined by

(3.1) $$K(z, v, X) = \frac{g(H(X,v)v, X)}{g(X,X)g(v,v) - g(X,v)^2}, \qquad z \in V(M)$$

K is a homogeneous function of degree zero in v. If we substitute X by the vector $Y = aX + bv$ ($a, b \in \mathbf{R}$), K remains unchanged. Thus K does not depend on the choice of X in the 2-plane P. (M, g) is to be isotropic at $x = pz \in M$ (scalar curvature in Berwald's terminology) if K is independent of X. We say that (M, g) is isotropic if it is isotropic at all the points. In this case the curvature tensor H of the Berwald connection is defined by

(3.2). $$H(Y, v)\, v = K(z)\, [g(v, v)\, Y - g(Y, v)v], \quad z \in V(M)$$

With respect to a local chart (x^i) of M, (3.2) becomes

(3.3) $$H^i_{\ ojo} = F^2\, K(z) h^i_{\ j}\,,\ h^i_{\ j} = (\delta^i_{\ j} - u^i u_j), \ \|u\| = 1$$

where δ denotes the Kronecker symbol. But the curvature tensor $H^i_{\ jkl}$ is obtained from $H^i_{\ 0k0}$ by

(3.4) $$H^i_{jkl} = \frac{1}{3}\,\partial_j(\partial_l H^i_{oko} - \partial_k H^i_{olo}) \quad (\partial_j = \frac{\partial}{\partial v^j})$$

On taking into account (3.3) we obtain

(3.5)
$$H^i_{jrl} = K(\delta^i_r g_{jl} - \delta^i_l g_{jr}) + (\delta^i_r v_l - \delta^i_l v_r)K_j + \frac{1}{3}(2v_j\delta^i_r - \delta^i_j v_r - v^i g_{jr})K_l$$
$$- \frac{1}{3}(2v_j\delta^i_l - \delta^i_j v_l - v^i g_{jl})K_r + \frac{1}{3}F^2(h^i_r K_{jl} - h^i_l K_{jr})$$

where $K_j = \partial_j K$, $K_{ij} = \partial^2_{ji} K$ from (3.5) we have the Ricci tensor $H_{ij} = H^r_{irj}$,

(3.6) $$H_{ij} = (n-1)(K(z)g_{ij} + v_j K_i) + \frac{1}{3}(n-2)(2v_i K_j + F^2 K_{ij})$$

If (M, g) is Riemannian H_{ij} is symmetric, by (3.6) it follows immediately that K_i is zero. On using the Bianchi identity we obtain for $n > 2$ that K = constant. Thus an isotropic Riemannian manifold ($n > 2$) in the above sense is of constant sectional curvature.

4. Properties of curvature tensors in the isotropic case

By (1.8) the curvature tensor R of the Finslerian manifold is related to the curvature tensor H of the Berwald connection by

(4.1). $$R^i_{jkl} = H^i_{jkl} + T^i_{jr}R^r_{0kl} + \nabla_l\nabla_0 T^i_{jk}$$
$$- \nabla_k\nabla_0 T^i_{jl} + \nabla_0 T^r_{jk}\nabla_0 T^i_{rl} - \nabla_0 T^i_{rk}\nabla_0 T^r_{jl}$$

where $\nabla_0 = v^i\nabla_i$

On multiplying the two sides of this relation by v^j and on taking into account (3.5), we obtain

(4.2). $R^i_{\ orl} = H^i_{\ orl} = K(\delta^i_{\ r}\, v_l - \delta^i_{\ l}\, v_r) + \frac{1}{3}F^2(h^i_{\ r}\, K_l - h^i_{\ l}\, K_r)$, $\quad (K_l = \dot\partial_l K)$

From the relation of commutation ([8.27, Chap II],)

$$R_{lijk} - R_{jkli} = T_{ijr}\ R^r_{\ 0lk} + T_{kir}\, R^r_{\ 0jl} + T_{jlr}\, R^r_{\ 0ki} + T_{lkr} R^r_{\ oij}$$

So, in virtue of (4.2)

(4.3) $\qquad R_{lijk} = R_{jkli}$

From the Bianchi identity (see Chap II) we obtain

(4.4) $\qquad \sigma_{(m,k,l)}\ \nabla_m R^i_{\ jkl} + \sigma_{(m,k,l)}\ Q^i_{\ jrm}\ R^r_{\ 0kl} + \sigma_{(m,k,l)}\ \nabla_k \nabla_o Q^i_{\ jml} = 0$

where σ denotes the sum of terms obtained by permuting cyclically the indices m, k and l. On multiplying the two sides of (4.4) by v^k and on taking into account (4.2) and (4.3) we then have in the isotropic case

(4.5)
$$\begin{aligned} R_{ijrl} &= K(g_{ir}g_{jl} - g_{rj}g_{il}) + kF^2\ Q_{ijrl} + \frac{1}{2}\nabla_o \nabla_o\ Q_{ijrl} \\ &+ \frac{1}{2}\ (g_{jl}\, v_r - g_{jr} v_l)K_i + \frac{1}{2}\ (g_{ri} v_l - g_{il} v_r)K_j \\ &+ \frac{1}{2}(g_{ir} v_j - g_{jr} v_i)K_l + \frac{1}{2}(g_{jl} v_i - g_{li} v_j)\ K_r \\ &+ \frac{1}{6}F^2[h_{jl}\, \dot\nabla_r K_i + h_{ir}\ \dot\nabla_l\ K_j - h_{jr}\ \dot\nabla_l K_i - h_{il}\, \dot\nabla_r K_j] \end{aligned}$$

where $\dot\nabla_r$ denotes the vertical covariant derivation. It follows that in the isotropic case the Ricci tensor $R_{\ jl} = R^r_{\ jrl}$ is symmetric. By (8.15 Chap II), the first two Bianchi identities containing the curvature tensor R reduce in the isotropic case to

(4.6) $\qquad \sigma_{(j,k,l)}\ R^i_{\ jkl} = 0$

(4.7) $\qquad \sigma_{(m,r,l)}\ \dot\nabla_m R^r_{\ jlk} + \sigma_{(m,r,l)}\ (F^2 \frac{1}{3} K_m + K v_m)\ \nabla_o Q^i_{\ jrl} = 0$

where $K_m = \dot{\partial}_m K$.

Finally in the isotropic case, from the Bianchi identities (see chap II) we obtain for the curvature tensors of the Berwald connection the identities

(4.8) $\sigma_{(J,k,l)} H^{i}_{jkl} = 0$

(4.9) $\sigma_{(m,k,l)} D_m H^{i}_{jkl} = 0$

(4.10) $D_l G^{i}_{jkm} - D_K G^{i}_{jlm} + \dot{\partial}_m H^{i}_{jkl} = \mathbf{0}$

II Finslerian manifolds with constant sectional curvature

1.Generalization of Schur's theorem.

A. CASE OF BERWALD CONNECTION

1) Berwald studies the Finslerian manifolds that satisfy

a) (M, g) is isotropic
b) K is independent of v.

By (3.5), we then have

(1.1) $H^{i}_{jkl} = K\,(\delta^{i}_{k} g_{jl} - \delta^{i}_{l} g_{jk})$

On contracting i and k in the identity (4.9.I) and on multiplying the two sides of the relation so obtained by v^j and v^l successively we obtain

(1.2) $D_i H^{i}_{olo} + D_0 H_{ol} - D_l H_{oo} = 0$

On taking into account (1.1) the above relation becomes for $n \neq 2$

$$(1.3) \qquad v_l D_0 K = F^2 D_l K$$

whence, by vertical derivation

$$g_{jl}\, D_0K + v_l D_j K = 2\, v_j\, D_l K$$

On multiplying the two sides by g^{jl} we get

$$D_0 K = 0$$

From the relation (1.3) it follows that K = constant ($n > 2$). Thus *an isotropic Finslerian manifold ($n > 2$), with K independent of the direction, is of constant sectional curvature in the Berwald connection.*

Now we modify Berwald's definition and introduce a function defined on a suitable fibre bundle.

2. Let G_2 (M) be the Grassmannian fibre bundle of 2-planes on M. We note by $\pi^{-1}G_2(M) \to W(M)$ the fibre induced on W(M) by π: $W(M) \to M$. Let $P \in \pi^{-1}G_2$ (M) a 2-plane generated by the vectors X and Y linearly independent at $x = \pi y \in M$.

Definition. *The sectional curvature at a point $y \in W(M)$ following the 2-plane P defined by the vectors X and Y at $x = \pi y \in M$ in the Berwald connection is defined to be the function $K : \pi^{-1} G_2 (M) \to R$*

$$(1.4) \qquad K_1(y, X, Y) = \frac{g(A(X,Y)Y,X)}{\|X\|^2\|Y\|^2 - g(X,Y)^2}$$

where A is the anti-symmetric tensor of H, defined by

$$(1.5) \qquad g(AX, Y)Y, X) = \frac{1}{2}[g(H(X,Y)Y,X) - g(H(X,Y)X,Y)]$$

$K_1(y, X, Y)$ is a homogeneous function of degree zero in v and remains unchanged if one replaces X and Y by the vectors

$X_1 = aX + bY$, $Y_1 = cX + dY$, with ad-bc $\neq 0$, a, b, c, d, $\in R$. Thus K_l does not depend on the choice of the vectors X and Y in the plane P. let suppose that K_l is independent of the plane P, then K_l is a fortiori independent of the plane passing the canonical vector v.

Now, $K_l(y, v, X) = K(y,v,X)$. Therefore (M, g) is isotropic (scalar curvature) in the Berwald sense. On the other hand, from (4.1.I) we have

(1.6). $$A_{ijkl} = R_{ijkl} + \nabla_0 T_{irk} \nabla_0 T^r{}_{jl} - \nabla_0 T_{jrk} \nabla_0 T^r{}_{il}$$

One easily verifies that in the isotropic case, in virtue of (4.3.1) and (4.6.1) the tensor A satisfies

(1.7)

1° $A_{ijkl} = - A_{jikl}$

2°. $A_{ijkl} = - A_{ijlk}$

3°. $\sigma_{(J,k,l)} A_{ijkl} = 0$

4°. $A_{ijkl} = A_{klij}$

Now K_l is independent of the plane and A satisfies the relations 1°, 2°, 3°, 4°. So by a classical reasoning ([23], tome I, p.200, proposition 1.4) that A must be of the form

$$A_{ijrl} = K\,(g_{ir}g_{jl} - g_{il}g_{jr})$$

Whence on multiplying the two sides by v^j

$$A_{iorl} = H_{iorl} = K\,(v_l g_{ir} - v_r g_{il})$$

(M, g) being isotropic in virtue of (4.2.I) we obtain

$$h^i{}_r K_l = h^i{}_l K_r\,, \quad h^i{}_r = \delta^i{}_r - u^i u_r \quad (\|u\| = 1)$$

On contracting i and r, we have, for n > 2, $K_1 = 0$. Thus K is independent of the direction. By a reasoning identical to the above case one finds that K is a constant.

Theorem. – *For K_1 (y, P) to be independent of the 2-plane P (dim M > 2) it is necessary and sufficient that curvature tensor H of the Berwald connection be defined by (1.1) where $K = K_1$ is an absolute constant[4].*

In this case, after (4.5.I), the curvature tensor R of the Finslerian connection is defined by

$$R_{ijkl} = K\,(g_{ik}g_{jl} - g_{jk}g_{il}) + K\,F^2\,Q_{ijkl} + \frac{1}{2}\nabla_0\nabla_0 Q_{ijkl} \tag{1.8}$$

B. CASE OF THE FINSLERIAN CONNECTION

By means of the curvature tensor R of the Finslerian connection one defines the function $K_2 : \pi^{-1}G(M) \to \mathbf{R}$

$$K_2(y, X, Y) = \frac{g(R(X,Y)Y,X)}{\|X\|^2\|Y\|^2 - g(X,Y)^2} \tag{1.9}$$

K_2 will be called the sectional curvature at $y \in W(M)$ following the 2-plane P(X, Y) generated by X and Y in the Finslerian connection. We have

$$K_2\,(y, v, X) = K(y, v, X) \tag{1.10}$$

As in the preceding case one proves

Theorem. [1] *For $K_2(y, P)$ to be independent of the two plane P(X, Y) (dim M > 2) it is necessary and sufficient that the curvature tensor R of the Finslerian connection be defined by*

$$R(X,Y)Z = K[g(Y,Z)\,X - g(X,Z)Y] \tag{1.11}$$

where K is a constant and X, Y, $Z \in T_x(M)$[4].

One shows that in this case for $K \neq 0$ the second curvature tensor P_{ijkl} is symmetric with respect to the last two indices k and l and the third curvature tensor of the Finslerian connection vanishes everywhere.

(1.12). $$Q = 0.$$

Such a manifold is a fortiori of constant sectional curvature in the Berwald connection. The converse is true if the condition (1.12) is satisfied.

2. Necessary and sufficient conditions for an isotropic Finslerian manifold to be of constant sectional curvature.

A. With the help of the Ricci curvature H_{ij} or R_{ij} one defines a scalar-valued function on W(M) by

$$\rho = H(u, u) = H_{ij}(x, u)\, u^i u^j = R_{ij}(x, u)\, u^i u^j. \; (u = 1).$$

We call ρ *the Ricci directional curvature.*

Theorem [4] – *For an isotropic Finslerian manifold to be of constant sectional curvature in the Berwald connection (n > 2), it is necessary and sufficient that the Ricci directional curvature H(u, u) satisfy the condition*

(2.1). $$\nabla_0 H(u, u) = 0$$

Proof. – By the identity (4.10.I) one obtains, on multiplying the two sides by v^l

(2.2). $$D_0 G^i_{jlm} + v^r \partial_m H^i_{jlr} = 0$$

(M, g) being isotropic, in virtue of (3.5.I), the above relation becomes

(2.3). $$D_0 G^i_{jlm} = 2 Kv^i T_{jlm} - \sigma_{(j, l, m)} \frac{1}{3} (\delta^i_l v_j + \delta^i_j v_l - 2v^i g_{jl}) K_m - \sigma_{(j, l, m)} \frac{1}{3} (F^2 \delta^i_l - v^i v_l) K_{jm}$$

where $\sigma_{(j, l, m)}$ denotes the sum of the terms obtained on permuting cyclically the indices j, i, and m, $K_m = \dot{\partial}_m K$, $K_{jm} = \dot{\partial}^2_{jm} K$. Let us contract i and l in (2.3)

(2.4). $$D_0 G_{jm} = -\frac{(n+1)}{3} (v_j K_m + v_m K_j + F^2 K_{jm}), \quad (G_{jm} = G^i_{jim})$$

Let us multiply the two sides of (2.4) by g^{jm}. We obtain

(2.5) $$F^2 g^{ij} \frac{\partial^2 K}{\partial v^i \partial v^j} + \frac{3}{(n+1)} D_0 G = 0, \quad (G = g^{jm} G_{jm})$$

(M, g) being isotropic, we have $(n-1)K = \rho$. Now by hypothesis ρ satisfies the condition (2.1). From (2.5) we obtain

(2.6). $$F^2 g^{ij} K_i K_j = \frac{1}{2} F^2 g^{ij} \frac{\partial^2 K^2}{\partial v^i \partial v^j} + \frac{3}{(n+1)} D_0 (KG)$$

The right hand side of (2.6) is a divergence on W(M). If M is compact, without boundary, by (2.6) it follows by integration over W(M)

$$\int_{W(M)} F^2 g^{ij} K_i K_j \eta(g) = 0$$

where $\eta(g)$ is the volume element of W(M). Hence $\partial_i K = 0$. Thus K is independent of the direction. So (M, g) is of constant sectional curvature in the Berwald connection. We will extend this result to the non-compact case. From the Bianchi identity we have obtained the relation (1.2). Now (M, g) is isotropic from (4.2.I). So we obtain, on contracting the indices i and r

(2.7) $$H_{0l} = (n-1)\, K(z)\, v_l + \frac{(n-2)}{3}\, F^2 K_l, \quad (K_l = \partial_l K)$$

Thus (1.2) becomes ($n \neq 2$)

$$F^2 D_l K = v_l D_0 K + \frac{1}{3} F^2 D_0 K_l$$

In virtue of (2.1), the above relation reduces to

(2.8) $$D_l K = \frac{1}{3} D_0 K_l$$

On the other hand, let us derive vertically the relation (2.1)

$$D_l K + D_0 K_l = 0$$

Now, on taking into account (2.8)

(2.9) $$D_l K = 0 = D_0 K_l$$

By vertical derivation, we obtain successively

(2.10) $$D_l (H_{oj} + H_{jo}) = 0$$

$$2 D_l \widetilde{H}_{jm} = (H_{or} + H_{ro}) G^r{}_{jlm}$$

where we have put

(2.11). $$\widetilde{H}_{jm} = \frac{1}{2} \frac{\partial^2 (H_{kl} v^k v^l)}{\partial v^j \partial v^m}$$

Now the curvature tensor $G^r{}_{jlm}$ is symmetric with respect to the indices j, l, and m. So from (2.10) we get

(2.12) $$D_l \widetilde{H}_{jm} = D_j \widetilde{H}_{ml} = D_m \widetilde{H}_{ij} = \widetilde{H}_{or} G^r{}_{jlm}$$

Let us write down the Ricci identity for the tensor $\widetilde{H}_{jm}$

$$D_k D_l \widetilde{H}_{jm} - D_l D_k \widetilde{H}_{jm} = - \widetilde{H}_{rm} H^r{}_{jkl} - \widetilde{H}_{jr} H^r{}_{mkl} - \partial_{\dot r} \widetilde{H}_{jm} H^r{}_{okl}$$

Let us multiply the two sides this relation by v^j. Then on taking into account (2.10), we have

$$\widetilde{H}_{rm} H^r{}_{okl} + \widetilde{H}_{or} H^r{}_{mkl} = 0$$

On multiplying this relation by v^l and then by g^{mk}, we have

$$\text{(2.13)} \qquad H^r{}_o{}^m{}_o \widetilde{H}_{rm} - H^r{}_{mok}\, g^{mk} \widetilde{H}_{or} = 0$$

Now from the expression $\rho = F^{-2} H_{ij} v^i v^j$, we have by vertical derivation and on taking into account (2.2)

$$2\widetilde{H}_{or} = H_{or} + H_{ro} = 2\, v_r\, \rho + F^2\, \partial_{\dot r}\, \rho$$

$$2\widetilde{H}_{mr} = 2\, \rho g_{mr} + 2\, v_m\, \partial_{\dot r}\, \rho + 2 v_r\, \partial_m\, \rho + F^2\, \partial^2_{\dot m \dot r}\, \rho$$

Thus from (2.13) we have

$$\text{(2.14)} \qquad D(\rho) = H^r{}_o{}^m{}_o \frac{\partial^2 \rho}{\partial v^r \partial v^m} - H^r{}_{mok}\, g^{mk} \frac{\partial \rho}{\partial v^r} = 0$$

(M, g) being isotropic, on calculating $H^r{}_o{}^m{}_o$ and $H^r{}_{mok}$ by (3.5.I), $D(\rho)$ becomes

$$\text{(2.15)} \qquad 2(n-1)\, D(\rho) = F^2 g^{ij} \frac{\partial^2 \rho^2}{\partial v^i \partial v^j} = 0, \qquad (\rho = (n-1)K)$$

Let us denote once again by ρ the restriction of ρ to the indicatrix S_x. we get

$$\partial_\alpha \rho^2 = F v^i{}_\alpha\, \partial_{\dot i}\, \rho^2$$

$$\partial_\beta \partial_\alpha \rho^2 = F \partial_\beta v'^i_\alpha\, \partial_i \rho^2 + F^2\, v'^i_\alpha\, v'^j_\beta\, \partial_j\, \partial_i \rho^2 + F\, v'^j_\beta\, \partial_j\, F v'^i_\alpha\, \partial_i \rho^2$$

Now $v'^i_\beta \partial_j F = 0$. So

$$\partial_\beta \partial_\alpha \rho^2 = F \partial_\beta\, v'^i_\alpha\, \partial_i \rho^2 + F^2\, v'^i_\alpha\, v'^j_\beta \partial_j\, \partial_i \rho^2$$

On taking into account (2.7.I) and on multiplying the two sides by $g^{\alpha\beta}$ we get

$$g^{\alpha\beta}\, \partial_\beta \partial_\alpha \rho^2 = F^2\, g^{ij}\, \partial_i\, \partial_j \rho^2 + F\, g^{\alpha\beta}\, \partial_\beta v'^i_\alpha \partial_i \rho^2$$

In virtue of (2.15.I), (2.15) becomes on S_x

(2.16) $D(\rho) = g^{\alpha\beta}\, \dot{\nabla}_\beta\, \partial_\alpha \rho^2 + A^\alpha\, \partial_\alpha \rho^2 = 0, \quad (\alpha, \beta = 1, 2, \ldots n\text{-}1)$

where $A^\alpha = g^{\lambda\beta}\, A^\alpha{}_{\lambda\beta}$. Now the indicatrix S_x is compact, it follows from the preceding equation using the maximum principle of Hopf that ρ^2 is constant on S_x. From (2.9) it then follows that K is an absolute constant.

B. Let us suppose that (M, g) is of constant sectional curvature in the Berwald connection. From (2.5) it follows that the second scalar curvature G must satisfy

(2.17) $\qquad D_o\, G = 0, \qquad (G = g^{jl} G^i{}_{jil})$

where $D_o = v^i\, D_i$. Conversely let suppose that (M, g) is isotropic and besides the scalar curvature G satisfies (2.17). From (2.5) we obtain

(2.17) $\qquad F^2\, g^{ij}\, \dfrac{\partial^2 K}{\partial v^i \partial v^j} = 0$

On restricting to the indicatrix S_x, we prove as in the preceding case that the equation (2.18) on S becomes

$$g^{\alpha\beta}\ \dot{\nabla}_{\beta}\partial_{\alpha}K + A^{\alpha}\partial_{\alpha}K = 0 \quad (\alpha,\beta = 1, \ldots n\text{-}1)$$

where we have denoted by K the restriction of K to S. By a reasoning analogous to the preceding case, we obtain

Theorem. [4] – *For an isotropic Finslerian manifold M (dim M > 2) to be of constant sectional curvature in the Berwald connection it is necessary and sufficient that second scalar curvature G satisfies (2.17).*

If dim M = 2, then (M, g) is isotropic and the tensor Q vanishes. If the scalar curvature G satisfies (2.17), then K is independent of the direction, and the tensors H and R become

$$R^{i}_{jkl} = H^{i}_{jkl} = K\,(\delta^{i}_{k}g_{jl} - \delta^{i}_{l}g_{jk})$$

where K is not necessarily a constant.

3. Locally Minkowskian manifolds

Let $\gamma : [0, b] \to M$ be a Finslerian geodesic, parametrized along its arc length on [0, b], γ is defined by a differential equation of the second order

$$\text{(3.1)} \qquad \frac{d^2x^i}{ds^2} + G^{i}_{jk}(x(s),\ \dot{x}(s))\ \frac{dx^j}{ds}\frac{dx^k}{ds} = 0,\ (\dot{x} = \frac{dx}{ds})$$

where G^{i}_{jk} are the coefficients of the Berwald connection.

We know that the data for

$$s = 0,\ x^{i}_{0} = x^{i}|_{s=0} \text{ and } \frac{dx^i}{ds}|_{s=0} = a^{i}$$

determine a unique geodesic starting from $x_o \in M$ tangent to the vector $A(a^i)$ at x_o and situated in a sufficiently small neighborhood of x_o. This geodesic is represented in the chart by [4]

(3.2) $$x^i = f^i(x_o^j, y^j), \quad y^j = a^j s$$

From (3.2) we obtain

$$\frac{dx^i}{ds} = \frac{\partial f}{\partial y^j}\frac{dy^j}{ds} = \frac{\partial f^i}{\partial y^j} a^j$$

whence, at the origin x_o

$$\frac{\partial f^i}{\partial y^j}\Big|_{s=0} = \delta^i_j$$

Thus the Jacobian of (3.2) is different from zero at the origin x_o. The equation is invertible in the neighborhood of x_o. This can be considered as change of local coordinates. In the new chart U(y) the geodesic γ is represented by the linear equation $y = as$. A second derivation gives us

$$\lim_{y\to 0}\frac{\partial^2 f^i}{\partial y^j \partial y^k} = -\lim_{y\to 0} G^i_{jk}(x_o, y)$$

Now the right hand side is indeterminate as $y \to 0$, since it is homogeneous of degree zero in y. Thus *the map* $y \to x$, *defined by (3.2) is a* C^1*–diffeomorphism of an open set of the zero section of TM over an open set of M. We call the coordinates (y) the normal geodesic coordinates.*

Theorem [4]. – *For the map f to be class* C^2 *it is necessary and sufficient that the second curvature tensor of the Berwald connection vanishes*

(3.3). $$G^i_{jkl} = 0 \Leftrightarrow \nabla_k T_{ijl} = 0$$

Definition. – *A Finslerian manifold is called locally Minkowskian if there exists a local chart* $U(x^i)$ *such that the fundamental function F defined on* $p^{-1}U(x^i, v^i)$ *does not depend on x [13].*

For a Finslerian manifold to be locally Minkowskian it is necessary and sufficient that two curvature tensors H and G of the Berwald connection vanish. This condition is equivalent to R = 0, and G = 0. Let us suppose that (M, g) is isotropic and in addition the tensor G vanishes. From the preceding considerations it follows that (M, g) is of constant sectional curvature. From (2.3) we have either (M, g) is Riemannian or K = 0. Thus, we have

Theorem. *An isotropic Finslerian manifold admitting normal geodesic coordinates of class C^2 is either Riemannian or locally Minkowskian[4].*

In dimension 2, this result is due to Busemann [11].

4. Compact isotropic manifolds with strictly negative curvature.

Let us suppose that (M, g) is isotropic. Let us multiply the two sides of (2.2) by v_i. We then have

$$(4.1).\qquad D_oD_oT_{jlm} + KFT_{jlm} + \frac{1}{3}F^2(h_{jl}K_m + h_{lm}K_j + h_{jm}K_l) = 0$$

with $h_{ij} = g_{ij} - u_i u_j$, $(u_i = F^{-1} v_i,\ K_j = \partial_j K)$

If K is independent of the direction, we obtain

$$(4.2).\qquad D_oD_o T_{jlm} + KF^2 T_{jlm} = 0$$

whence

$$(4.3).\qquad \frac{1}{2} D_oD_o\, g(T, T) + KF^2 g(T, T) - g(D_oT, D_oT) = 0$$

Let us suppose that M is compact without boundary and K < 0. Then the first term of (4.3) is a divergence on W(M). On integrating over W(M) we obtain $D_oT = 0$. From (4.2) we have T = 0. Thus (M, g) is Riemannian.

More generally

Theorem. – *A compact without boundary isotropic Finslerian manifold (M, g) (dim M > 3), with symmetric indicatrix and with strictly negative sectional curvature is a Riemannian manifold with constant sectional curvature.*

Proof. For $x = x_o$ fixed, let $g_{\alpha\beta}(v(t)) = g_{\alpha\beta}(x_o, v(t))$ the metric of the indicatrix S_o defined by (2.5I). Let $\hat{g}_{\alpha\beta}$ be a metric conformal to $g_{\alpha\beta}$

(4.4) $$\hat{g}_{\alpha\beta} = e^{2\sigma(v(t))} g_{\alpha\beta} \quad \alpha, \beta = 1, \ldots (n-1)$$

We denote by $\hat{D}$ the Riemannian covariant derivative associated to $\hat{g}_{\alpha\beta}$, we have

(4.5) $$\hat{D}_\beta v^i_\alpha = \dot{\nabla}_\beta v^i_\alpha - (\delta^\mu_\beta \sigma_\alpha + \delta^\mu_\alpha \sigma_\beta - g_{\alpha\beta}\sigma^\mu) v^i_\mu, \quad \left(\sigma_\alpha = \frac{\partial \sigma}{\partial t^\alpha}\right)$$

where $\dot{\nabla}_\beta v^i_\alpha$ is defined by (2.15.I). We call pseudo-normal the vector of components

(4.6). $$N^i = \frac{1}{n-1} g^{\alpha\beta} \dot{\nabla}_\beta v^i_\alpha = -\frac{1}{n-1} A^\lambda v^i_\lambda - v^i \qquad (A^\lambda = g^{\alpha\beta} A^\lambda_{\alpha\beta})$$

It is clear that the (M, g) is Riemannian if and only if N is normal to S (A = 0) [21]. Let us put

(4.7). $$Z^i = \frac{1}{n-1} \hat{g}^{\alpha\beta} \hat{D}_\beta v^i_\alpha$$

Let us choose

(4.8). $$\sigma_\sigma = -\frac{1}{(n+1)} A_\alpha \qquad (A_\alpha = A^\lambda_{\alpha\lambda})$$

(4.9) $$e^{2\sigma} Z^i = 2\sigma^\lambda v^i_\lambda - v^i$$

(4.10) $$e^{2\sigma} \dot{\nabla}_\alpha Z^i = a^\lambda_\alpha v^i_\lambda$$

(4.11) $$\hat{D}_\beta v^i_\alpha = C^\lambda_{\alpha\beta} v^i_\lambda + \hat{g}_{\alpha\beta} Z^i$$

With

(4.12). $a^{\lambda}{}_{\alpha} = 2(\dot{\nabla}_{\alpha}\sigma^{\lambda} - 2\sigma^{\lambda}\sigma_{\alpha} - \sigma^{\mu} A^{\lambda}{}_{\mu\alpha}) - \delta^{\lambda}{}_{\alpha}$

(4.13) $C^{\lambda}{}_{\alpha\beta} = -[A^{\lambda}{}_{\alpha\beta} - \frac{1}{n-1}(\delta^{\lambda}{}_{\beta}A_{\alpha} + \delta^{\lambda}{}_{\alpha}A_{\beta} + g_{\alpha\beta} A^{\lambda})]$

Now, M is isotropic; by (4.1) it follows, on eliminating the vertical derivative of K

(4.14) $D_o D_o \tau_{jlm} + KF^2 \tau_{jlm} = 0$

with

(4.15) $\tau_{jlm} = T_{jlm} - (1/(n+1))(h_{jl}T_m + h_{lm}T_j + h_{mj}T_l) \quad (T_m = T^i{}_{im})$

On multiplying the two sides of (4.14) by τ^{jlm} we obtain

(4.16) $\frac{1}{2} D_o D_o \|\tau\|^2 = \|D_o\tau\|^2 - KL^2\|\tau\|^2$

where $\|\tau\|^2$ is the square of the norm of τ. The left hand side of (4.16) is a divergence on W(M) [7.9 Chapter III]. M supposed to be compact without boundary, we have by integrating over W(M)

$$\int_{W(M)} [\|D_o\tau\|^2 - KF^2\|\tau\|^2] = 0$$

Now K < 0, Then $D_o\,\tau = 0$. From (4.14) it follows that

(4.17) $\tau_{jlm} = 0$

On projecting this relation on the tangent space to the indicatrix S_{xo} we obtain $C^{\lambda}{}_{\alpha\beta} = 0$. Thus (4.10) and (4.11) become

(4.18) $e^{2\sigma}\dot{\nabla}_{\alpha}Z^i = [2(\dot{\nabla}_{\alpha}\sigma^{\lambda} + \delta^{\lambda}{}_{\alpha}\sigma_{\mu}\sigma^{\mu}) - \delta^{\lambda}{}_{\alpha}]v^i{}_{\lambda}$

(4.19) $\hat{D}_{\beta}v^i{}_{\alpha} = \hat{g}_{\alpha\beta} Z^i$

From (4.19) it follows immediately by derivation, on taking into account (4.18) and on using the properties Riemann curvature tensor $\hat{R}$ associated to $\hat{D}$

(4.20) $$\hat{R}^{\lambda}{}_{\alpha\gamma\beta} = \psi(\hat{g}_{\alpha\beta}\delta^{\lambda}{}_{\gamma} - \hat{g}_{\alpha\gamma}\delta^{\lambda}{}_{\beta})$$

(4.21) $$\dot{\nabla}_{\beta}\sigma_{\alpha} + g_{\alpha\beta}\,\sigma_{\mu}\sigma^{\mu} = \chi\, g_{\alpha\beta} \qquad \psi = e^{-2\sigma}(1 - \chi)$$

After (4.20) the indicatrix Sx_o $(n > 3)$, equipped with the metric $\hat{g}_{\alpha\beta}$ is of constant sectional curvature and we have

(4.22) $\dot{\nabla}_{\alpha}Z^{i} = -\psi\,\dot{\nabla}_{\alpha}v^{i}$, $\quad \psi = \text{constant}$

By integrating we obtain

(4.23) $$Z^{i} = -\psi\, v^{i} + c^{i}, \quad \dot{\nabla}_{\alpha} c^{i} = 0$$

where c^{i} is a constant vector. Now the indicatrix Sx_o being supposed to be symmetric, that is to say $F(x_o, -v) = F(x_o, v)$, by (4.23) it follows that on changing v into –v and on taking into account of the expression Z defined by (4.9) that Z is transformed into –Z. Consequently, $Z^{i} = -\psi\, v^{i}$. Thus Sx_o is a sphere and (M, g) is a Riemannian manifold with constant sectional curvature

III. Complete manifolds with constant sectional curvatures

1. Operator D^1. The Isotropic Case

We suppose that (M, g) is isotropic. For every function f: W(M) →R, we put

(1.1). $$D^{1}f = D_{\hat{u}}\, D_{\hat{u}}\, D_{\hat{u}} f + 4K\, D_{\hat{u}} f + 2\, D_{\hat{u}} K.f$$

where $\hat{u} = F^{-1}\hat{v}$ is horizontal over u $(\rho\hat{u} = u)$. If f and ψ are two functions on W(M) we have

$$g(D^{1}f, \psi) + g(f, D^{1}\psi) = \text{div on } W(M)$$

From this we deduce that if M is compact, D^1 is *anti-auto-adjoint.* Thanks to the tensor τ, defined by (4.15.II) we construct the tensor

$$(1.2).\qquad S^i{}_{jkl} = \tau^i{}_{lr}\tau^r{}_{jk} - \tau^i{}_{kr}\tau^r{}_{jl}$$

(M, g) being isotropic, on taking into account the relation (4.14.II) we obtain

$$(1.3).\qquad D^1\,(F^2 S^i{}_{jkl}) = 0$$

It follows, by contraction $S = g^{jl}S^i{}_{jil}$

$$D^1(L^2 S) = 0$$

With

$$(1.4).\qquad S = \|\tau\|^2 = \|T\|^2 - \frac{3}{(n+1)}\|T^*\|^2$$

where T* denotes the torsion trace vector. We note that the tensor T majors the tensors τ and T*. The relation (1.3) shows that the kernel of D^1 is non-empty. Let Q be the third curvature tensor of the Finslerian connection [See Chapter II]

$$(1.5).\qquad Q^i{}_{jkl} = T^i{}_{lr}T^r{}_{jk} - T^i{}_{kr}T^r{}_{jl}$$

Let us denote by M the tensor

$$(1.6).\qquad M^r{}_{ijkl} = F(h_{ik}T^r{}_{jl} + h_{jl}T^r{}_{ik} - h_{il}T^r{}_{jk} - h_{jk}T^r{}_{il})$$

with $h_{ij} = g_{ij} - u_i u_j$, $u = F^{-1}\,v$

In the isotropic case, we obtain

$$(1.7).\qquad D^1(F^2 Q_{ijkl}) = D_o M^r{}_{ijkl}\,K_r + \tfrac{1}{3} M^r{}_{ijkl}\,D_o\,K_r$$

where $K_r = \partial_r K$. The tensor S defined by (1.2) is related to Q by

(1.8). $$S_{ijkl} = Q_{ijkl} + h_{il}\, T_{jk} + h_{jk} T_{il} - h_{ik}\, T_{jl} - h_{jl}\, T_{ik}$$

with

(1.9). $$T_{jk} = \frac{1}{(n+1)^2}\left[T_j T_k + \frac{1}{2} h_{jk} T_r T^r - (n+1) T_{jkr}\, T^r\right]$$

From (1.4) it follows

(1.10). $$\|\tau\|^2 = Q + \frac{(n-2)}{(n+1)} \|T^*\|^2 = \frac{3}{(n+1)} Q + \frac{(n-2)}{(n+1)} \|T\|^2$$

2. Complete manifolds with strictly negative constant sectional curvature

Let us suppose that (M, g) is isotropic and $n > 2$. We have shown in paragraph (2.II) that if $D_{\hat{u}} K$ is zero everywhere, then the manifold is of constant sectional curvature. The operator D^1, defined by (1.1) becomes

(2.1). $$D^2 f = D_{\hat{u}}\, D_{\hat{u}}\, D_{\hat{u}}\, f + 4K\, D_{\hat{u}}\, f, \qquad (K = \text{constant})$$

It is clear that the tensor $F^2\, S^i{}_{jkl}$ is in the kernel of this operator. In the same way are the tensors, by (1.7), $F^2 Q^i{}_{jkl}$ and the contracted tensors $F^2 Q_{jl}$, $F^2\, Q$

(2.2). $$D^2(F^2\, S) = 0, \quad D^2(F^2 Q) = 0$$

In virtue of (1.10) it follows that

(2.3) $$D^2\, \|FT^*\|^2 = 0, \quad D^2\, \|FT\|^2 = 0$$

Let us rewrite $(2.3)_1$

(2.4) $$D_{\hat{u}}\, D_{\hat{u}}\, D_{\hat{u}}\, F^2 \|T^*\|^2 + 4K\, D_{\hat{u}}\, F^2 \|T^*\|^2 = 0$$

Let us put

$$D_{\hat{u}} L^2 \|T^*\|^2 = f, \; 4K = - C^2 \tag{2.5}$$

Now γ is a geodesic on M, parametrized according to its arc length s, the equation (2.4) along γ becomes

$$f'' - C^2 f = 0, \qquad (f'' = \frac{d^2 f}{ds^2}) \tag{2.6}$$

The roots of the characteristic polynomial are $\pm$ C, whence the general integral

$$f = C_1 e^{Cs} + C_2 e^{-Cs} \tag{2.7}$$

where C_1 and C_2 are arbitrary constants on γ .

In the following we suppose that (M, g) is *geodesically complete*. On choosing A and B such that

$$A = \pm 2\sqrt{C_1 C_2} \quad B = \frac{1}{2} \log \frac{C_1}{C_2} \tag{2.8}$$

f is put under the form

$$f = A \, ch \, (Cs + B)$$

Let us denote $\varphi = \|FT^*\|^2$. Along the geodesic γ

$$\varphi = \frac{A}{C} sh \, (Cs + B) + D$$

where D is a constant on γ. Let us suppose that the norm of the torsion vector is bounded on W(M). If $s \to \pm \infty$, $sh \, (Cs + B) \to \pm\infty$, and φ is bounded only if A =0. That is to say f = 0. By (2.5) one obtains along the geodesic γ

$$g(T^*, D_{\hat{u}} T^*) = 0$$

A second derivation gives us

$$g(T^*, D_{\hat{u}} D_{\hat{u}} T^*) + g(D_{\hat{u}} T^*, D_{\hat{u}} T^*) = 0$$

In virtue of (4.2.II), the above equation is reduced to

$$-Kg(T^*, T^*) + g(D_{\hat{u}} T^*, D_{\hat{u}} T^*) = 0$$

Now K is strictly negative, from the above relation it follows that T* vanishes along γ. But the geodesic γ is arbitrary. So T* vanishes everywhere. Hence (M, g) is Riemannian.

Theorem. – *A geodesically complete Finslerian manifold (dim M > 2) with strictly negative constant sectional curvature and with bounded torsion vector is Riemannian[4].*

3. Complete manifolds with strictly positive constant sectional curvature

In this paragraph, we suppose that the indicatrix S_x is symmetric

$$F(x, \lambda v) = |\lambda| F(x, v), \lambda \in \mathbf{R})$$

and (M, g) is of strictly positive constant sectional curvature. In addition we suppose that (M, g) is metrically complete. By the generalization of Myers' theorem, M is compact. The tensors $F^2 Q'^{i}_{jkl}$ and R'^{i}_{jkl} belong to the kernel of the operator D^2. We therefore have

$$\text{(3.1)} \qquad D^2 \overline{Q} = D_{\hat{u}} D_{\hat{u}} D_{\hat{u}} \overline{Q} + 4K D_{\hat{u}} \overline{Q} = 0$$

where $\overline{Q} = F^2 Q$. Let γ be a geodesic of M parametrized by its arc length s, from (3.1) one obtains along γ

(3.2) $$\frac{d^2\overline{\varphi}}{ds^2} + k^2\,\overline{\varphi} = 0, \qquad (k^2 = 4K)$$

where $\overline{\varphi} = \frac{d\overline{Q}}{ds}$. In the following we assume that $\overline{Q} = F^2.Q$ is independent of the direction. By (2.17.I), the indicatrix S_{x_0} for $x = x_0$ fixed has a constant scalar curvature. Thus $\overline{Q}$ is a function defined on M. The preceding equation determines a solution for $\overline{Q}$ along γ

(3.3) $$\overline{Q} = A.\cos ks + B.\sin ks + C$$

where A, B, and C are constants on γ. Now M is compact. So the function $\overline{Q}$ admits an absolute maximum and minimum on M. Let us suppose that this maximum is attained for s = 0 and that the value of $\overline{Q}$ at this point is + 1. Then we have

(3.4) $$\frac{d\overline{Q}}{ds}\Big|_{s=0} = kB = 0, \quad \overline{Q}\,(0) = A + C = 1$$

Let $P_0 \in M$, a point corresponding to s = 0. $\overline{Q}$ admits an absolute minimum for π/k. Let us denote by $P_1 \in M$ this point and let us suppose that the value of $\overline{Q}$ at this point is equal to –1. We have

(3.5) $$\overline{Q}\,(\pi/k) = -A + C = -1.$$

From (3.4) and (3.5) it follows that

(3.6) $$\overline{Q} = \cos ks$$

One proves (see [3] pp 67-71) that $\overline{Q}$ **admits only P_0 and P_1 as critical points.** If in addition one assumes that M is simply connected, then by a result of Milnor [30], M is homeomorphic to an n-sphere (n > 2).

Theorem. – *Let (M, g) (dim M > 2) be a simply connected Finslerian manifold, complete, with a symmetric indicatrix of strictly positive constant sectional curvature in the Berwald connection. If moreover the scalar curvature of the indicatrix at each point of the manifold is independent of the direction, then (M, g) is homeomorphic to an n-sphere[4].*

4. Complete manifolds with zero sectional curvature

If the sectional curvature in the Berwald connection vanishes, by (3.1.I) (3.4.I) and (1.4.II) the curvature tensor H vanishes everywhere. The equation $(2.3)_2$ becomes

$$D_{\hat{u}}\, D_{\hat{u}}\, D_{\hat{u}}\, \|FT\|^2 = 0 \tag{4.1}$$

If we denote by $\varphi = \|FT\|^2$ we have along a geodesic γ

(4.2). $$\varphi = as^2 + bs + c$$

where a, b, and c are constants on γ. Suppose that φ is bounded on W(M). From (4.2) one deduces that it can be so only if a= b = 0. Therefore φ is a constant on γ.

We have

$$D_{\hat{u}}\, \|FT\|^2 = 0 = F^2\, g(D_{\hat{u}}\, T, T) \tag{4.3}$$

On the other hand K being zero, by (4.1.II) it follows that the torsion tensor must satisfy

(4.4). $$D_{\hat{u}}\, D_{\hat{u}}\, FT = 0$$

Let us derive the relation (4.3) along γ, taking into account (4.4). We obtain

$$g(D_{\hat{u}}\,FT,\; D_{\hat{u}}\,FT) = 0$$

The geodesic γ being arbitrary in the above relation, it follows

(4.5). $$\nabla_0 T_{ijk} = 0 \Leftrightarrow P_{ijkl} = 0$$

Thus (M, g) is a Landsberg manifold. Now H is zero by (4.1.I). So it follows immediately that curvature tensor R of the Finslerian connection vanishes everywhere (R= P= 0)

Let us derive vertically the relation (4.5)

(4.6). $$\nabla_l T_{ijk} = -\nabla_0 \nabla_{\dot{l}}\, T_{ijk}$$

On multiplying the two sides by $\nabla^{l}\, T^{ijk}$ we obtain

$$\nabla^{l}\, T^{ijk} \nabla_l T_{ijk} = -\frac{1}{2}\, \nabla_0 (\nabla^{l}\, T^{ijk} \nabla_{l}\, T_{ijk})$$

(4.7). $$\nabla^{l}\, T^{ijk} \nabla_l T_{ijk} = \frac{1}{2} \nabla_0\, \nabla_0\, (\, \nabla^{l\,\bullet}\, T^{ijk}\, \nabla_{\bullet\, l}\, T_{ijk})$$

But the covariant derivative of the type $\nabla_0 = D_0$ the left hand side is zero. So we have

(4.8) $$D_{\hat{u}}\, D_{\hat{u}}\, D_{\hat{u}}\, \left\| F\dot{\nabla} T \right\|^2 = 0$$

where $\dot{\nabla}$T is the vertical covariant derivation of T. Let us now suppose that vertical covariant derivative of the torsion tensor is bounded on W(M). On integrating the equation (4.8) along a geodesic γ and on reasoning as above, we conclude that the geodesic γ and on reasoning as above, we conclude that the derivative of $\left\|F\dot{\nabla}T\right\|^2$ along the geodesic γ is zero. So the left hand side of (4.7) vanishes on γ. But γ is an arbitrary geodesic so that $\nabla_i T_{jkl}$ vanishes everywhere. Therefore (M, g) is locally Minkowskian.

Theorem. – *Let (M, g) be a Finslerian manifold (dim M > 2), geodesically complete with zero Berwald curvature. If the torsion tensor and its vertical covariant derivative are bounded over W(M) then (M, g) is locally Minkowskian[4].*
Corollary. – *All compact Finslerian manifolds (dim M > 2) with zero Berwald curvature are locally Minkoswkian[4].*

IV. The Plane Axioms in Finslerian Geometry

1.Finslerian submanifolds [7]

Let i : S → M be a k-dimensional submanifold of M. We identify a point x in S with its image i(x) and a tangent vector $X \in T_x(S)$ to its image i*(X) where i* is the linear tangent map. Thus $T_x(S)$ becomes a subspace of $T_x(M)$. The canonical imbedding i induces a map $\tilde{i}$: V(S) → V(M) where $z \in V(S)$ is identified to $\tilde{i}$ (z). Thus V(S) is fibre sub-bundle of V(M) and the restriction of p to V(S) will be denoted by q : V(S) → S ; we also denote by $\overline{T}$ (S) = i^{-1} T(M) the fibre bundle induced from T(M) by i. The Finslerian metric of M induces over S a Finslerian metric which we again denote by g. At a point $x = qz \in S$ $(z \in V(S))$, we denote by Nqz the orthogonal complement of Tqz(S) in $\overline{T}$ qz(S). ($\overline{T}$ x(S) = Tx(S) + Nx(S)), Tx (S)$\cap$ Nx (S) = 0. We denote by P_1: $\overline{T}$ (S) → T(S), and P_2 : $\overline{T}$ (S) → Nx the projections and we put

$$(1.1) \qquad q^{-1}\overline{T}(S) = q^{-1}T(S) \oplus N$$

where N is called the normal fibre and is identified canonically to

$q^{-1}\overline{T}(S)/q^{-1}T(S)$. If TV(S) is the tangent bundle to V(S) we denote again by ρ the canonical linear map of $TV(S)\to q^{-1}T(S)$. Let $\hat{X}$ and $\hat{Y}$ be two vector field on V(S). For $z \in V(S)$, $(\nabla_{\hat{X}} Y)z$ belongs to $\overline{T}_{qz}(S)$. We have by (1.1)

$$(1.2) \qquad \nabla_{\hat{X}} Y = \overline{\nabla}_{\hat{X}} Y + \alpha(\hat{X}, Y),\ Y = \rho(\hat{Y})$$

where ∇ is the covariant derivation in the Finslerian connection. From (1.2), it follows that $\overline{\nabla}$ is a covariant derivation in the fibre bundle $q^{-1}T(S) \to V(S)$ and is Euclidean ($\overline{\nabla} g = 0$). Now $\alpha(\hat{X}, \rho(\hat{Y}))$, having values in N, is bilinear in $\hat{X}$ and $\hat{Y}$. We have by (11.1 Chap I)

$$(1.3) \qquad \overline{\tau}(\hat{X}, \hat{Y}) = P_1\tau(\hat{X}, \hat{Y}) = \overline{\nabla}_{\hat{X}} Y - \overline{\nabla}_{\hat{Y}} X - \rho[\hat{X}, \hat{Y}]$$

$$(1.4) \qquad P_2\tau(\hat{X}, \hat{Y}) = \alpha(\hat{X}, Y) - \alpha(\hat{Y}, X),\ X = \rho(\hat{X}),\ Y = \rho(\hat{Y})$$

where $\overline{\tau}$ is the torsion of the connection $\overline{\nabla}$. Let $v : S \to V(S)$ the canonical section and $\hat{v}$ the vector field on V(S) over v. ($\rho\hat{v} = v$). From (1.2) we get

$$(1.5) \qquad \nabla_{\hat{X}} v = \overline{\nabla}_{\hat{X}} v + \alpha(\hat{X}, v)$$

For $x \in S$ the fibre $q^{-1}(x)$ is a submanifold of $p^{-1}(x)$. Every vertical tangent vector $V\hat{X}$ to $q^{-1}(x)$ at a point $z \in q^{-1}(x)$ can be identified to a vertical vector of V(M). From (1.5) we get

$$\nabla_{V\hat{X}} v = \overline{\nabla}_{V\hat{X}} v + \alpha(V\hat{X}, v)$$

After (1.4), on taking into account the homogeneity of the torsion tensor T we get

$$(1.6) \qquad P_2\tau(V\hat{X}, \hat{v}) = \alpha(V\hat{X}, v) = 0$$

We therefore have

(1.7) $$\nabla_{V\hat{X}}\, v = \overline{\nabla}_{V\hat{X}}\, v$$

Thus if $V_zV(S)$ is the vertical tangent space to $q^{-1}(x)$ at $z \in V(S)$ the restriction of the map μz to $V_zV(S)$ ($z \in V(S)$) defines an isomorphism of this space with $T_{qz}(S)$. On the other hand, the relation (1.5) proves that if $\hat{X}$ is a $\overline{\nabla}$-horizontal vector field on V(S), it does not belong to the horizontal distribution defined by ∇. Thus $\tau(h\hat{X}, h\hat{Y}) \neq 0$, where $h\hat{X}$ and $h\hat{Y}$ are $\overline{\nabla}$- horizontal We put

(1.8) $$\overline{T}(\dot{X}, Y) = \overline{\tau}(V\hat{X}, h\hat{Y}),\ \overline{S}(X, Y) = \overline{\tau}(h\hat{X}, h\hat{Y})$$

where $\dot{X} = \nabla_{V\hat{X}}\, v$, $X = \rho(\hat{X})$, $Y = \rho(\hat{Y})$. Finally if $\hat{X}$, $\hat{Y}$ and $\hat{Z}$ are three vector fields on V(S) projectable by q, by (1.1.I) we obtain

(1.9) $$g(\overline{T}(\dot{X}, Y), Z) = g(\overline{T}(\dot{X}, Z), Y).$$

The torsion tensor corresponding to the Finslerian case admits the same property of symmetry. We call $\overline{\nabla}$ the induced connection and $\alpha(\hat{X}, Y)$ the second fundamental form of the submanifold S. To the induced metric we can associate also a Finslerian connection $\dot{\nabla}$ which will be called the *intrinsic connection* of the submanifold. *In order that the two connections* $\dot{\nabla}$ *and* $\overline{\nabla}$ *coincide it is necessary and sufficient that the tensor* $\overline{S}$ *is zero.*

Let $x \in S$ and a $\overline{U}(x)$ a neighbourhood of M containing x, endowed with local coordinates (x^{α}, x^{β}) $(\alpha, \beta = 1 \ldots k,\ a, b = k+1 \ldots n)$. Let U be a neighbourhood of x such that

$$U = \{P \in \overline{U}(x) \mid x^{k+1}(P) = \ldots = x^{n}(P) = 0\}$$

The $(x^{\alpha})_{|U}$ define a system of local coordinates of S. With respect to this system the induced connection $\overline{\nabla}$ and the 2-form $\alpha(\hat{X}, \hat{Y})$ become for $\hat{X}$, and $\hat{Y} \in TV(S)$,

$$(1.10) \qquad \overline{\nabla}_{\hat{X}}\left(\frac{\delta}{\delta x^{\beta}}\right) = \pi^{\alpha}_{\beta}(\hat{X})\,\frac{\delta}{\delta x^{\alpha}}, \qquad \pi^{\alpha}_{\beta} = \omega^{\alpha}_{\beta} - g^{\alpha c}k_{ca}\omega^{a}{}_{\beta}$$

$$(1.11) \qquad \alpha(\hat{X}, Y) = g^{ia}\,k_{ac}Y^{\beta}\,\omega^{c}{}_{\beta}(\hat{X})\,\frac{\delta}{\delta x^{i}}, \qquad Y = \rho(\hat{Y})$$

where k_{ac} is the inverse of g^{ab} and ω is the Finslerian connection of M. If we denote by ω^{λ}_{β} on $q^{-1}(U)$ the 1-form of the intrinsic Finslerian connection on S, we obtain after (1.7.**I**),

$$(1.12) \quad 2\overset{0}{G}_{\gamma} = \overset{0}{\Gamma}_{\gamma\alpha\beta}\, v^{\alpha}v^{\beta}$$
$$= \frac{1}{2} v^{\alpha}\, v^{\beta}\,(\delta_{\alpha}g_{\beta\gamma} + \delta_{\beta}g_{\alpha\gamma} - \delta_{\gamma}\, g_{\alpha\beta}), \quad (\overset{0}{\Gamma}_{\gamma\alpha\beta} = g_{\gamma\lambda}\overset{0}{\Gamma}{}^{\lambda}{}_{\alpha\beta})$$

The expression of the right hand side of (1.12) is identical to that coming from the ambient connection ∇ that is

$$(1.13) \qquad 2\overset{\circ}{G}_{\gamma} = 2g_{\gamma i}G^{i} = 2g_{\gamma i}\,(G^{i} - g^{ic}k_{ca}G^{a})$$

since $g_{\gamma i}\, g^{ic} = 0$. After (1.10) we put

$$(1.14) \qquad \overline{G}^{i} = G^{i} - g^{ic}\, k_{ca}G^{a}$$

Now $\overline{G}^{b} = 0$, the relation (1.13) becomes

$$(1.15) \qquad \overset{\circ}{G}_{\gamma} = g_{\gamma\lambda} \overset{\circ}{G}{}^{\lambda} = g_{\gamma\lambda} \overline{G}^{\lambda}$$

From this we conclude that $\overset{\circ}{G}{}^{\lambda} = \overline{G}^{\lambda}$. *Thus the induced connection and the intrinsic connection define the same geodesic.* From the relation (1.12) we obtain, by vertical derivation, on taking into account (1.15) and the expression of the coefficients of the intrinsic connection

$$(1.16) \qquad \overset{0}{\Gamma}{}^{\lambda}{}_{\sigma\mu}\, v^{\sigma} = \delta^{\bullet}_{\mu} \overline{G}^{\lambda}$$

At the same time the analogous expression which is used in the calculation of the coefficients of the induced connection is, after (1.10),

$$(1.17) \qquad \pi^{\lambda}{}_{\sigma\mu}\, v^{\sigma} = \delta^{\bullet}_{\mu} G^{\lambda} - g^{\lambda c}\, k_{ca}\, \delta^{\bullet}_{\mu} G^{a}$$

We thus have

$$(1.18) \qquad \overline{\nabla} V^{\lambda} = \nabla V^{\lambda} - (g^{\lambda a} k_{ac})\, \delta^{\bullet}_{\mu} . G^{c}\, dx^{\mu}$$

2. Induced and Intrinsic Connections of Berwald

Let S be a k-dimensional submanifold of M, $\hat{X}$ and $\hat{Y}$ be two vector fields on V(S) and D the Berwald connection. By the decomposition of the tangent space at the point $x = qz \in S$, (1.1),

$(D_{\hat{X}}\, Y)_x$ becomes

$$(2.1) \qquad D_{\hat{X}}\, Y = \overline{D}_{\hat{X}}\, Y + \beta(X, Y), \;\; X = \rho(\hat{X}), \; Y = \rho(\hat{Y})$$

We deduce from it that $\overline{D}$ is a covariant derivation in the vector bundle $q^{-1}T(S) \to V(S)$. It is without torsion and satisfies

$\overline{D}_{\hat{v}}g = 0$ where $\hat{v}$ is horizontal ($\overline{D}_{\hat{v}}v = 0$) over v, ($\rho(\hat{v}) = v$). Now β has values in **N** being bilinear symmetric in X and Y. From (1.5) and (2.1) it follows

$$(2.2) \qquad \overline{\nabla}_{\hat{X}}v = \overline{D}_{\hat{X}}v, \; \alpha(X, v) = \beta(X, v), \qquad X = \rho(\hat{X})$$

Thus the two induced connections of Finsler and Berwald define the same splitting in the tangent space to V(S). It thus follows the geodesics of two induced connections coincide with those of the intrinsic connection. If $\hat{X}$ is a $\overline{\nabla}$ - horizontal vector field over V(S), on taking into account (1.8.I) we obtain

$$(2.3) \qquad \overline{D}_{\hat{X}}Y = \overline{\nabla}_{\hat{X}}Y + P_1(\nabla_{\hat{v}}T)(X, Y)$$

$$(2.4) \qquad \beta(X,Y) = \alpha(X, Y) + P_2(\nabla_{\hat{v}}T)(X, Y), \; X = \rho(\hat{X})$$

With respect to a local chart $(x^\alpha)_{|U}$ (λ, α= 1, …k, a, b,= k+1,…n) adapted to the submanifold S we have

$$(2.5) \qquad \overline{D}\left(\frac{\delta}{\delta x^\lambda}\right) = \overline{B}^\alpha_\lambda \frac{\delta}{\delta x^\alpha}$$

$$\overline{B}^\alpha_\lambda = B^\alpha_\lambda - g^{\alpha a}k_{ab} B^b_\lambda = (G^\alpha{}_{\lambda\gamma} - g^{\alpha a}k_{ab}G^b{}_{\lambda\gamma})dx^\gamma$$

and

$$(2.6) \qquad \beta(X, Y) = g^{ia}\, k_{ab}\, X^\lambda\, B^b{}_\lambda\, (Y)\frac{\delta}{\delta x^i}$$

On the other hand the intrinsic connection of Berwald is obtained from $\overline{G}^\lambda = G^\lambda - g^{\lambda a}k_{ab}G^b$ by vertical derivation say $\dfrac{\delta^2 \overline{G}^\lambda}{\delta v^\alpha \delta v^\beta}$

3.TotallyGeodesic Submanifolds

Definition. – *A submanifold S of M is said to be geodesic at* $m \in S$ *if every geodesic starting from m of S is a geodesic of M. It is said to be totally geodesic if it is geodesic at all its points[4].*

Let us suppose that S is totally geodesic. Let γ: $[0,a] \to S$ be a geodesic of S, starting from $\gamma(0) = m$, parametrized by its curvilinear abscissa over [0,a], u the vector field of unit tangents to γ, $\hat{\gamma}$ its canonical horizontal lift in the unitary fibre bundle w(S) and $\hat{u}$ the vector field tangent along $\hat{\gamma}$ over u. We have after (1.5)

$$\text{(3.1)} \qquad \overline{\nabla}_{\hat{u}}\, u = 0 \Rightarrow \nabla_{\hat{u}}\, u = 0 \Rightarrow \alpha(\hat{u}, u) = 0$$

Let us consider the class of curves defined by $\hat{\gamma}$ in V(S). An element of this class is a curve above $\hat{\gamma}$, let $(x(s), v(s) = \lambda u(s))$, $\lambda > 0$ and $s \in [0, a]$. We then have since γ is a geodesic

$$\nabla_{\hat{u}}\, v = i(\hat{u})d\lambda . u(s)$$

where i(,) denotes the interior product. Since u is unitary we have

$$g(u, \nabla_{\hat{u}}\, v) = i(\hat{u})\, d\lambda$$

whence

$$\nabla_{\hat{u}}\, v = g(u, \nabla_{\hat{u}}\, v) . u(s).$$

Thus $\hat{u}$ considered as a vector in the space TV(S) becomes

$$\text{(3.2)} \qquad \hat{u} = u^a \frac{\partial}{\partial x^a} + g(u, \nabla_{\tilde{u}}\, v)\, u^a \frac{\partial}{\partial v^a}$$

After (1.11), the condition $\alpha(\hat{u}, u) = 0$ becomes

$$2G^a = \Gamma^{*a}_{\beta\lambda}(x, v)\, v^\beta v^\lambda = 0$$

This relation is valid whatever be the geodesic γ, whence

(3.3)
$$\frac{\partial G^a}{\partial v^\lambda} = \Gamma^{*a}{}_{\beta\lambda}\,(x, v)\, v^\beta = 0.$$

We thus have after (1.11) and (1.6)

(3.4)
$$\alpha(\hat{X}, v) = 0$$

whatever be the vector field $\hat{X}$ on V(S). Conversely if (3.4) is satisfied for every $\hat{X} \in TV(S)$ then S is totally geodesic. Let us note that after (2.6) this condition is equivalent to $\beta(X, Y) = 0$, whatever be X and Y over S.

Theorem. *In order that a submanifold S of a Finslerian manifold M be totally geodesic it is necessary and sufficient that the second fundamental form satisfies (3.4) whatever be the vector field $\hat{X}$ over V(S)[4].*

From the above theorem it follows that for a totally geodesic submanifold the induced connections from Finsler or from Berwald coincide with the corresponding intrinsic connection.

4. The Plane Axioms

Axiom 1.---*Let (M,g) a Finslerian manifold of dimension ≥ 3. For every point $x \in M$ and every subspace E_2 of dimension two of $T_x(M)$ there exists a surface S passing through x, totally geodesic such that $T_x(S) = E_2$* [4]

Let (M, g) be Finslerian manifold, dim $M \geq 3$, satisfying the axiom 1. Let $y = (x, u)$ be any element of the unitary fibre bundle W(M) where u is the unitary tangent vector to M at $x = \pi y$ where $\pi : W(M) \to M$. Let Y and Z be two orthonormal vectors at x and orthogonal to u

$$g(Z,Y) = g(Z, u) = g(Y, u) = 0, \quad \|Z\| = \|Y\| = \|u\| = 1.$$

By hypothesis there exists a surface S, passing through x, totally geodesic and tangent to the plane defined by u and Y. Let $\hat{Y}$ and $\hat{u}$ be two vector fields horizontal on V(S) over Y and u, then we have , after (11.2.I Chapter I)

$$(4.1) \qquad R(Y,u)\,u = \nabla_{\hat{Y}} \nabla_{\hat{u}} u - \nabla_{\hat{u}} \nabla_{\hat{Y}} u - \nabla_{[\hat{Y},u]} u = - \nabla_{[\hat{Y},\hat{u}]} u$$

S being totally geodesic, we have

$$(4.2) \qquad \nabla_{[\hat{Y},\hat{u}]} u = \overline{\nabla}_{[\hat{Y},\hat{u}]} u$$

where $\overline{\nabla}$ is the connection induced over S. By (4.1) we have

$$g(R(u,Y)Z,u) = g(R(Y,u)u, Z) = - g(\overline{\nabla}_{[\hat{Y},\hat{u}]} u, Z) = 0$$

This relation is valid whatever be the vector fields Y and Z orthonormal and orthogonal to u. If one chooses Y' and Z' in the plane defined by Y and Z such that

$$(4.3) \qquad Y' = \frac{1}{\sqrt{2}}(Y + Z), \qquad Z' = \frac{1}{\sqrt{2}}(Y - Z),$$

It is clear that Y′ and Z′ are orthonormal and orthogonal to u we have

$$0 = g(R(u,Y')Z',u) = g(R(u,Y)Y,u) - g(R(u,Z)Z,u)$$

The expression $K_u = g(R(u, Y)Y, u)$ is therefore a function of (x, u) and does not depend on the plane passing through u. It thus follows that (M, g) is isotropic. Thus *every Finslerian manifold satisfying the axiom 1 is isotropic.*

Let now $x \in M$ and X, Y, Z three orthonormal vectors at x. Let us denote once again by S the surface totally geodesic passing through x tangent to the plane defined by X and Y. Let v the canonical section S $\rightarrow$ V(S) Let us denote by $\hat{X}$ and $\hat{Y}$ two vector fields horizontal above X and Y.

Since $g(Z,X) = 0$, we have by covariant derivation

(4.5) $$g(\nabla_{\hat{Y}} Z, -(\nabla_0 T)(Y,Z), X) = 0$$

Similarly, since $g(Z,Y)= 0$, we have

(4.5) $$g(\nabla_{\hat{Y}} Z - \nabla_0 (T)(Y,Z), Y) = 0$$

Thus $\nabla_{\hat{Y}} Z - (\nabla_0 T)(Y,Z)$ is orthogonal to the plane defined by X and Y. From (4.5) we have by covariant derivation

(4.6) $$g(\nabla_{\hat{X}} \nabla_{\hat{Y}} Z - \nabla_{\hat{X}} ((\nabla_0 T)(Y,Z)), X)$$
$$= g(\nabla_{\hat{Y}} Z, (\nabla_0 T)(X, X)) - g((\nabla_0 T)(Y, Z),(\nabla_0 T)(X,X))$$

By the same reasoning we obtain

(4.7) $$g(\nabla_{\hat{Y}} \nabla_{\hat{X}} Z - \nabla_{\hat{Y}} ((\nabla_0 T)(X, Z)), X)$$
$$= g(\nabla_{\hat{X}} Z, (\nabla_0 T)(X, Y)) - g((\nabla_0 T)(X,Z),(\nabla_0 T)(X, Y)).$$

Let us derive the relation (4.4)

(4.8) $$(\nabla_{[\hat{X},\hat{Y}]} Z, X) + g(Z, \nabla_{[\hat{X},\hat{Y}]} X) = 0$$

Since S is totally geodesic $\beta = 0$, we have after (1.8.I)

(4.9) $$\nabla_{H[\hat{X},\hat{Y}]} X = \overline{D}_{H[\hat{X},\hat{Y}]} X - (\nabla_0 T)(\nabla_{\hat{X}} Y - \nabla_{\hat{Y}} X, X)$$

On the other hand $\hat{X}$ and $\hat{Y}$ are horizontal; so we have

(4.11) $$\nabla_{[\hat{X},\hat{Y}]} v = - R(X, Y) v$$

In virtue of (1.4) and the above relation we obtain

(4.12) $\nabla_{V[\hat{X},\hat{Y}]} X = \overline{\nabla}_{V[\hat{X},\hat{Y}]} X - P_2 T(R(X, Y)v, X)$

On taking into account (4.10) and (4.12) the relation (4.9) becomes

(4.13)
$g(\nabla_{[\hat{X},\hat{Y}]} Z, X) = g(R(X, Y)v, T(Z, X)) + g(\nabla_{\hat{X}} Y - \nabla_{\hat{Y}} X, (\nabla_0 T)(Z, X))$

After (1.6.II) we put

(4.14)
$$g(A(X, Y)Z, X) = g(R(X, Y)Z, X) + g((\nabla_0 T)(X, X)\,(\nabla_0 T)(Y, Z)) - g((\nabla_0 T)(X, Z), (\nabla_0 T(X, Y))$$

Subtracting the sum of the relations (4.13), (4.8) from the relation (4.7) and taking into account (4.14) we obtain

(4.15) $g((A(X, Y)Z, X) = g((\nabla_{\hat{X}} \nabla_0 T)(X, Z), Y) - g((\nabla_{\hat{Y}} \nabla_0 T)(X, Z), X) - g(R(X, Y)v, T(Z, X))$

Now (M, g) is isotropic, so in virtue of the properties of the tensor A, (1.7.II), g(A(X, Y) Z, X) is symmetric with respect to Y and Z. Hence also the right hand side. Thus (4.15) becomes

(4.15)′ $g(A(X, Y)Z, X) = g((\nabla_{\hat{X}} \nabla_0 T)(X, Z), Y) - g((\nabla_{\hat{Z}} \nabla_0 T)(X, Y), X) - g(R(X, Y)v, T(Y, X))$

Let us put Y = u with g(Z, u) = g(X, u) = g(Z, X) = 0. In virtue of the homogeneity of the tensor T the right hand side of (4.15)′ is zero. Hence $g(A(X, u)Z, X) = 0$. On the other hand we have $g(A(X, X)Z, X) = 0$. Now $Y = au + bX$. Thus we obtain

$$g(A(X,Y)Z,X) = 0$$

Let us make the transformation defined by (4.3), we get

$$0 = g(A(X, Y)\,Y, X) - g(A(X, Z)Z, X)$$

It thus follows that $K_1 = g(A(X,Y)Y,X)$ depends only on (x, u). By the generalization of the theorem of Schur (1.II) K_1 is a constant.

Theorem. *Every Finslerian manifold satisfying axiom 1 is of constant sectional curvature in the connection of Berwald.*

Definition. *A submanifold S of a Finslerian manifold M is semi-parallel if for every vector field* $\hat{X}$ $\overline{\nabla}$*-horizontal over V(S) we have for all Y*

$$\alpha(\hat{X}, Y) = 0 \tag{4.16}$$

From the relation $(2.2)_2$ it follows that S is totally geodesic and we have

$$\beta(X,Y) = 0, \quad P_2(\nabla_0 T)\,(X,Y) = 0 \tag{4.16$'$}$$

where X and Y are tangent to S.

Axiom 2. *Let (M, g) be a Finslerian manifold (dim $M \geq 3$). For every point $x \in M$, and for every subspace E_2 of dimension two if there exists a surface passing through x, semi-parallel such that $T_x(S) = E_2$.*

Suppose (M, g) satisfies axiom 2. Let X,Y, Z be three orthonormal vectors at $x = pz$, $z \in V(M)$.

$$g(X, Y) = g(Y, Z) = g(Z, X) = 0, \quad \|X\| = \|Y\| = \|Z\| = 1.$$

We denote by S the surface passing through x and tangent to the plane defined by X and Y. Let $\hat{X}$ and $\hat{Y}$ be two vector fields horizontal on V(S) over X and Y respectively. Since S is semi-parallel we have

$$g(\nabla_{\hat{Y}} Z, X) = 0 \quad g(\nabla_{\hat{Y}} Z, Y) = 0$$

Thus $\nabla_{\hat{Y}} Z$ is orthogonal to the plane (X,Y). So also is $\nabla_{\hat{X}} Z$. From the above relations, we obtain, by covariant derivation,

$$(4.17) \qquad g(\nabla_{\hat{X}} \nabla_{\hat{Y}} Z, X) = 0, \; g(\nabla_{\hat{Y}} \nabla_{\hat{X}} Z, X) = 0$$

On the other hand by (4.12) we have

$$(4.18) \qquad g(\nabla_{[\hat{X},\hat{Y}]} Z, X) = g(R(X,Y)v, T(Z,X))$$

From (4.17) and (4.18) we get

$$(4.19) \qquad g(R(X,Y)Z,X) + g(R(X, Y)\, v, T(Z,X)) = 0$$

where v is the canonical section of $S \to V(S)$. But (M, g) has constant sectional curvature in the Berwald connection. So we have

$$R(X,Y)v = K(g(Y, v)X - g(X,v)Y)$$

where K is a constant. Thus (4.19) can be written

(4.20)

$$g(R(X,Y)Z,X) + Kg(Y, v).g(T(Z,X),X) - Kg(X,v).g(T(Z,X),Y) = 0$$

The first term g(R(X,Y)Z,X) is symmetric with respect to Y and Z. So we have

$$Kg(Y, v).g(T(Z,X),X) = Kg(Z, v).g(T(Y,X), X) = 0$$

On the other hand in the tangent plane $T_x(S)$, Y has the form Y = av + bX. Thus the last term of (4.20) is equally zero. So we have

$$(4.21) \qquad g(R(X,Y)Z,X) = 0$$

By an identical reasoning we also obtain

$$(4.22) \qquad g(R(X,Y)Y,X) = g(R(X,Z)Z,X)$$

Thus the common value of (4.22) does not depend on the plane (X,Y). It is a function of (x, v). By the generalization of Schur's theorem (paragraph 1.**II**) it is a constant.

Theorem. *Every Finslerian manifold satisfying axiom 2 has constant sectional curvature in the Finslerian connection*[4].
Definition. *A submanifold S of a Finslerian manifold (M, g) is called auto-parallel if the second fundamental form of S in the Finslerian connection is everywhere zero.*

$$(4.23) \qquad \alpha(\hat{X},Y) = 0, \;\; Y = \rho(\hat{Y})$$

where X and Y are two vector fields over V(S).

Axiome 3. *Let (M, g) be a Finslerian manifold (dim M $\geq$ 3) For every point x $\in$ M et every subspace E_2 of dimension two of $T_x(M)$ there exists a surface S, autoparallel passing through x such that $T_x(S) = E_2$.*

Let us now suppose that (M, g) satisfies axiom 3. Let X, Y, Z be three orthonormal vectors at x $\in$ M; there exists a surface S passing through x, tangent to the plane (X,Y) and auto-parallel. If $\hat{X}$ and $\hat{Y}$ are two vector fields on V(S) over X and Y, we find by an identical reasoning to the preceding case, and taking into account of (4.23), the relation

(4.24) $$g(\Omega(\widehat{X},\widehat{Y})Z,X) = 0$$

where Ω is the curvature two form. We choose $\widehat{X}$ and $\widehat{Y}$ such that

(4.25) $$\nabla_{V\hat{X}} v = X, \quad \nabla_{V\hat{Y}} v = Y$$

where v is the canonical section of $S \to V(S)$. From (4.24) we obtain

(4.26) $$g(Q(X,Y)Z,X) = 0$$

But the curvature tensor Q has the same symmetry property as the curvature tensor R in the isotropic case. On making the transformation (4.3) we obtain

(4.27) $$g(Q(X,Y)Y,X) = g(Q(X,Z)Z,X)$$

Thus the common value of (4.27) depends only on (x, u). We deduce from it easily that Q = 0 (see [1], pp.46-47)

If one chooses in (4.24), $\widehat{X}$ and $\widehat{Y}$ horizontal above X and Y, we obtain the relation (4.21). This proves that (M, g) has the constant sectional curvature in the Finslerian connection. We have to consider the case where $\widehat{X}$ is horizontal and $\widehat{Y}$ vertical so that $\rho(\widehat{X}) = X$, $\nabla\widehat{Y}\, v = Y$

We obtain

(4.28) $$g(P(X,Y)Z, X) = 0$$

where the left hand side is defined by (see [1] p.36)

(4.29) $$\begin{aligned} g(P(X, Y)Z, X) &= g((\nabla_{\hat{X}} T)(Y,Z), X) \\ &-g((\nabla_{\hat{Z}} T)(X, Y), X) + g(T(X, (\nabla_o T)(Y,Z), X) \\ &-g(T(X, Z), (\nabla_o T)(X, Y)) \end{aligned}$$

where $\hat{Z}$ is horizontal over Z. The relation (4.28) then can be written

(4.30)
$$g((\nabla_{\hat{X}} T)(Y, Z), X)) + g(T(X, (\nabla_o T)(Y, Z)), X))$$
$$= g((\nabla_{\hat{Z}} T)(X,Y), X) + g(T(X, Z),(\nabla_o T)(X, Y))$$

where the left hand side is symmetric with respect to Y and Z. Hence it must be so for the right hand side also

(4.31)
$$g((\nabla_{\hat{Z}} T)(X, Y), X) + g(T(X, Z), (\nabla_0 T)(X, Y))$$
$$= g((\nabla_{\hat{Y}} T)(X, Z), X) + g(T(X, Y), (\nabla_0 T)(X, Z))$$

where $\hat{Y}$ is horizontal over Y. Now let us make the transformation (4.3). Then on taking into account (4.31) we obtain

(4.32)
$$g(P(X, Z)Y, X) = g(P(X, Y)Z, X)$$

Thus the common value of this expression is independent of the plane. On choosing Y = u, we conclude that this common value is zero. From this we deduce easily (see [1], p.46) that the tensor P is identically zero. If the sectional curvature is non zero then by (4.2.II) (M, g) is a Riemannian manifold of constant sectional curvature. If the sectional curvature is zero, it follows from the above argument that the curvature two form Ω is everywhere zero.

Theorem. *Let (M,g) be a Finslerian manifold satisfying axiom 3. Then either(M, g) is a Riemannian manifold with constant sectional curvature or the two form Ω of Finslerian curvature of (M, g) is everywhere zero*[4].

CHAPTER VII

PROJECTIVE VECTOR FIELDS ON THE UNITARY TANGENT FIBRE BUNDLE [3]

(**abstract**) We give a characterization of the projective vector fields on Finslerian unitary tangent fibre bundle and we prove that in the compact case the existence of projective vectors is related to the sign of the flag curvature. We define the notion of a restricted infinitesimal projective transformation, and introduce a new projective invariant tensor. We determine the necessary and sufficient conditions for the vanishing of projective invariants. We study the case when the Ricci directional curvature (R.D.C) is satisfied under certain conditions especially when the curvature is constant. (This is a generalization of Einstein manifold). We show that any simply connected, metrically complete, Finslerian manifold with a R.D.C. positive constant and admitting a proper vector field leaving the covector of torsion trace invariant is homeomorphic to an n-sphere.

1. Infinitesimal Projective Transformations

Let X be a vector field over $U \subset M$; for u sufficiently small it defines a local 1-parameter group denoted by exp(uX) of local transformations of U.

We denote by $\exp(u\hat{X})$ its extension to $p^{-1}(U)$ and we have

$p \circ \exp(u\hat{X}) = \exp(uX) \circ p$ (see chapter III). With respect to a frame adapted to the decomposition defined by the Finslerian connection $\hat{X}$ becomes at $z \in p^{-1}(U)$:

$$(1.1) \qquad \hat{X} = X^i \frac{\partial}{\partial x^i} + \nabla_0 X^i \frac{\partial}{\partial v^i},$$

where $\frac{\partial}{\partial x^i}$ and $\frac{\partial}{\partial v^i}$ are Pfaffian derivations with respect to $(dx^i, \theta^i = \nabla v^i)$, and $\nabla_0 = \nabla_{\hat{v}}$. From (11.1 Chapter I), and for a Finslerian connection it follows that we have $\nabla_{\hat{X}} v = \nabla_{\hat{v}} X$. Let

∇v be the *1-form of splitting* associated to the canonical vector field v. Its Lie derivative is

$$(1.2) \qquad L(\hat{X})\nabla v(\hat{Y}) = L(\hat{X})\nabla_{\hat{Y}} v - \nabla_{[\hat{X},\hat{Y}]} v$$

for any vector field $\hat{Y}$ on V(M). If $\hat{Y}$ is vertical we see easily that the right hand side is zero (see 5.8 chapter III); then from the structure equation we obtain:

$$(1.3) \quad L(\hat{X})\nabla v(H\hat{Y}) = \nabla_{H\hat{Y}} \nabla_{\hat{v}} X + R(X, Y)v + (\nabla_{\hat{v}} T)(Y, \nabla_{\hat{v}} X).$$

where $H\hat{Y}$ is the horizontal part of $\hat{Y}$.

Definition . A vector field X over M is a projective infinitesimal transformation for the Finslerain connection if

$$(1.4) \qquad L(\hat{X})\nabla v(H\hat{Y}) = \Psi_*(v).Y + \Psi_*(Y)v.$$

where $\hat{Y}$ is a vector field over V(M) such that :

$$(1.5) \qquad \nabla_{\hat{Y}} v = \rho(\hat{Y}) = Y, \qquad \nabla_{V\hat{Y}} \Psi = \Psi_*(Y)$$

and Ψ is a homogeneous function of degree one at v over V(M), Ψ_* denotes the vertical differential of Ψ ($\Psi_*(Y) = Y^i \dot{\partial}_i \Psi$)

Equating (1.3) to (1.4), we get

$$(1.6) \quad \nabla_{H\hat{Y}} \nabla_{\hat{v}} X + R(X, Y)v + (\nabla_{\hat{v}} T)(Y, \nabla_{\hat{v}} X) = \Psi_*(v)Y + \Psi_*(Y)v.$$

Let us put in (1.6), $H\hat{Y} = \hat{v}$, $(\nabla_{\hat{v}} v = 0)$; on taking into account the homogeneity of T :

$$(1.7) \qquad \nabla_{\hat{v}} \nabla_{\hat{v}} X + R(X, V)v = 2\,\Psi_*(v)v.$$

Conversely if X satisfies (1.7), by covariant vertical derivation we prove that X is a infinitesimal projective transformation. Let us suppose that X is an infinitesimal projective transformation , the function Ψ is defined, from (1.7), by

$$2\Psi = F.\, g(\nabla_{\hat{u}}\, \nabla_{\hat{u}} X, u), \tag{1.8}$$

where $u = F^{-1}.v$ and $\hat{u} = F^{-1}\hat{v}$ To every vector field X on M, we associate the vector field $\hat{Z}$ on V(M) defined by :

$$\hat{Z} = \hat{X} - g(X, u)\,\hat{u} - g(\nabla_{\hat{u}} X, u)\,\hat{v}\ , \tag{1.9}$$

where $\hat{v} = v^i \frac{\delta}{\delta v^i}$, whence

$$Z = \rho(\hat{Z}) = X - g(X, u)u. \tag{1.10}$$

If X is an infinitesimal projective transformation then Z satisfies

$$\nabla_{\hat{v}}\, \nabla_{\hat{v}} Z + R(Z, v)v = 0 \tag{1.11}$$

and conversely. Let us express the relation (1.11) in terms of the bracket $[\hat{Z}, \hat{v}]$. For this we have at first :

$$\nabla_{\hat{v}} Z = \nabla_{\hat{v}} X - g(\nabla_{\hat{u}} X, u)v = \nabla_{\hat{X}} v - g(\nabla_{\hat{u}} X, u)\ \nabla_{\hat{v}} v = \nabla_{\hat{Z}} v$$

Now from the first structure equation (11.1 Chapter I), for a Finslerian connection it follows that $[\hat{Z}, \hat{v}]$ is vertical and taking into account the second structure equation for the Finslerian connection, (1.11) becomes:

$$\nabla_{\hat{v}}\, \nabla_{\hat{v}} Z + R(Z, v)v = -\ \nabla_{[\hat{Z}, \hat{V}]}\, v = 0$$

Thus the bracket $[\hat{Z}, \hat{v}]$ is horizontal ; it is therefore identically zero.

The equation (1.11) with $\hat{Z}$ defined by (1.9) is therefore equivalent to the invariance of horizontal vector fields $\hat{v}$ by $\hat{Z}$.

Proposition 1. *In order that a vector field X over M defines a projective infinitesimal transformation for the Finslerian connection it is necessary and sufficient that the vector field $\hat{Z}$ associated to X by (1.9) leaves invariant the horizontal vector field $\hat{v}$ [3].*

2. Other characterizations of infinitesimal projective transformations.

To every vector field X on M, we associate on the one hand the symmetric covariant tensor of order two defined by:

$$(2.1)\ t(Y, Z) = g(\nabla_{H\hat{Y}} X, Z) + g(\nabla_{H\hat{Z}} X, Y) + 2g\,(T(\nabla_{\hat{v}} X, Y), Z)$$

where $\hat{Y}$ and $\hat{Z}$ are vector fields on V(M) ; on the other hand, the vector field I(X), section of $p^{-1}T(M)$ defined by :

$$(2.2) \qquad I(X) = \nabla_{\hat{v}} \nabla_{\hat{v}} X + R(X, v)v.$$

We note that t is none other than the Lie derivative of the metric tensor by X.

Lemma *1. We have the formula* :

$$(2.3) \qquad (\nabla_{\hat{v}} t)(Y, v) - \frac{1}{2}(\nabla_{H\hat{Y}} t)(v, v) = g(I(X), Y).$$

Proof. By definition we have

$$(2.4) \qquad (\nabla_{\hat{v}} t)\,(Y, v) = \nabla_{\hat{v}}(t(Y, v)) - t(\nabla_{\hat{v}} Y, v)\,.$$

when $\hat{v}$ is horizontal over v ($\rho\hat{v} = v$)

From (2.1) we obtain:

(2.5) $$t(Y, v) = g(\nabla_{H\hat{Y}} X, v) + g(\nabla_{\hat{v}} X, Y),$$

(2.6) $$t(v,v) = 2g\,(\nabla_{\hat{v}} X, v).$$

On calculating the left hand side of (2.3) by (2.4) (2.5) and (2.6) as well as the structure equations we obtain the formula.

Lemma 2. *Let X be a vector field on M and a function ψ defined on V(M) by*

(2.7) $$g(\nabla_{\hat{v}}\, \nabla_{\hat{v}} X, v) = 2\,\Psi g(v, v),$$

we then have for every Y satisfying (2.5) :

(2.8) $$2(\nabla_{\hat{v}}\, t)\,(Y, v) - g(I(X), Y) = 2\,\Psi_*(Y)\, F^2 + 4\,\psi\, g(Y, v)\,.$$

Proof. Let $\hat{Y}$ be a vector field satisfying (1.5) ; from (2.7) we obtain by vertical covariant derivation ($\hat{\dot{Y}} = V\hat{Y}$)

(2.9) $$g(\nabla_{\hat{\dot{Y}}}\, \nabla_{\hat{v}}\, \nabla_{\hat{v}} X, v) + g(\nabla_{\hat{v}}\, \nabla_{\hat{v}}\, X, Y) = 2\Psi_*(Y)\, F^2 + 4\,\Psi g(Y, v).$$

On the other hand, by the structure equations we have

$$\nabla_{\hat{\dot{Y}}}\, \nabla_{\hat{v}}\, \nabla_{\hat{v}} X = \nabla_{\hat{v}}\, \nabla_{\hat{\dot{Y}}}\, \nabla_{\hat{v}} X\, .+ \nabla_{[\hat{\dot{Y}},\hat{V}]}\, \nabla_{\hat{v}} X$$

Now

$$\rho\, [\hat{\dot{Y}}, \hat{v}] = Y, \qquad \nabla_{[\hat{\dot{Y}},\hat{v}]}\, v + \nabla_{\hat{v}}\, Y = 0$$

$$\nabla_{\hat{\dot{Y}}}\, \nabla_{\hat{v}} X = \nabla_{\hat{v}}\, \nabla_{\hat{\dot{Y}}} X + \nabla_{[\hat{\dot{Y}},\hat{v}]} X = (\nabla_{\hat{v}}\, T)(Y, X) + T(Y,\, \nabla_{\hat{v}} X) + \nabla_{H\hat{Y}} X$$

and

$$\nabla_{[\hat{\dot{Y}},\hat{v}]}\, \nabla_{\hat{v}} X = \nabla_{\hat{v}}\, \nabla_{[\hat{\dot{Y}},\hat{v}]} X + \nabla_{[[\hat{\dot{Y}},\hat{v}],\hat{v}]} X + R(Y, v)X$$

$$\rho[[\hat{Y},\hat{v}],\hat{v}] = -2\ \nabla_{\hat{v}} Y$$

Using the definition of t and the above formulas we obtain the lemma.

Proposition 2. *In order that a vector field X defines an infinitesimal projective transformations for the Finslerian connection it is necessary and sufficient that it satisfies one of the equivalent conditions* [3] *:*

(1) $$(\nabla_{\hat{v}} t)(Y, v) - \frac{1}{2}(\nabla_{H\hat{Y}} t)(v, v) = 2\,\Psi g(Y, v).$$

(2) $$(\nabla_{\hat{v}} t)(Y, V) = 3\,\Psi g(Y, v) + \Psi_*(Y)\, g(v, v).$$

(3) $$(\nabla_{\hat{v}} t)(Y,Z) = 2\,\Psi\, g(Y, Z) + \Psi_*(Z) g(Y, v) + \Psi_*(Y)\, g(Z, v).$$

Proof. The condition (1) follows from the proposition 1 and lemma 1. The condition (1) implies condition (2), since on putting Y = v in (1) we obtain the relation (2.7), in virtue of the lemma 2, we then have the condition (2). Conversely (2) implies the condition (1) as it suffices to put Y = v in (2). We then have (2.7) from lemma (2) and in virtue of the condition (2) we obtain (1). So (1) and (2) are equivalent and characterize the infinitesimal projective transformations. On the other hand (3) implies (2), it suffices to put Z = v in (3). To show (1) implies (3) we choose a local chart of M; the condition (1) then becomes in this chart:

(2.10) $$\nabla_o t_{io} = \frac{1}{2}\nabla_i t_{oo} + 2\Psi v_i$$

where $t_{ij} = L(\hat{X})\, g_{ij}$ is defined by (2.1). From (2.10) one obtains, by covariant vertical derivation and on taking into account the second identity of Ricci, ([Chapter I]) the relation

$$\nabla_j t_{io} + \nabla_o t_{ij} = \nabla_i t_{jo} + 2\,\Psi_{j'} v_i + 2\Psi g_{ij}$$

where $\Psi_i = \delta^{\cdot}_i \Psi$. On exchanging the indices i and j and on adding it to this relation we obtain

(2.11) $$\nabla_o t_{ij} = 2\Psi g_{ij} + v_i\Psi_j + v_j\,\Psi_i \quad (\Psi_i = \delta_i^{\cdot}\,\Psi)$$

This relation is none other than the condition (3).

3. Curvature and infinitesimal projective transformations [3]

Let $S_x : g_u(u,u) = 1$ be the indicatrix at x ;we have denoted by $W(M) = \bigcup_{x\in M} S_x$ the unitary tangent fibre bundle on M, by $\eta(g)$ the volume element of W(M). The divergences of the differential forms on W(M) are calculated relative to $\eta(g)$ (see Chapter III)

We have shown that if X is a projective vector field, we have (1.11) where Z is defined by (1.10)

Lemma 1. *Let (M, g) a compact Finslerian manifold without boundary; for every projective X we have :*

(3.1) $$\langle \nabla_{\hat{u}} Z, \nabla_{\hat{u}} Z\rangle = \int_{W(M)} g(R(X,u)u, X)\,\eta(g),$$

where $\langle,\rangle$ denotes the global scalar product and where $\rho(\hat{u})=u$, $\nabla_{\hat{u}}\, u = 0$. We note that $\nabla_{\hat{u}} Z$ is vertical part of the lift of X on W(M).

Proof. If X is projective, from (1.11) we have

$$\nabla_{\hat{u}}\, g(\nabla_{\hat{u}} Z, Z) = \frac{1}{2}\nabla_{\hat{u}}\nabla_{\hat{u}}\, g(Z, Z)$$
$$=(\nabla_{\hat{u}} Z, \nabla_{\hat{u}} Z) - g(R(X, u)u, X)\,.$$

The left hand side is a divergence on W(M), by integration we obtain the lemma.

Lemma 2. *Let (M, g) a compact Finslerian manifold without boundary; for every vector field X on M we have*

$$(3.2)\quad n\int_{W(M)} g(R(X,u)u,X)\,\eta(g) = \int_{W(M)} \Phi(X,X)\,\eta(g),$$

with

$$(3.3)\qquad \Phi(X,X) = [H_{ij} + 2(\nabla_j\nabla_o T_i - \nabla_r\nabla_o T^r{}_{ij})\,X^iX^j]$$

where ($H_{ij} = H^r{}_{irj}$) is the Ricci tensor of the Berwald connection

Proof. By the Bianchi identity written in local coordinates for a Berwald connection, we obtain the identity (see 4.10 I Chapter VI))

$$(3.4)\qquad \dot\delta_m H^i{}_{jkl} + D_l G^i{}_{jkm} - D_k G^i{}_{jlm} = 0, \qquad (\dot\delta_m = \frac{\partial}{\partial v^m})$$

whence, on multiplying the two sides by v^j ,

$$(3.5)\qquad v^j\,\dot\delta_m H^i{}_{jkl} = 0$$

On the other hand by (§4.4.1. I Chap VI) the curvature tensor R of the Finslerian connection is linked to H by

(3.6)

$$R_{ijkl} = H_{ijkl} + T_{ijr}R^r{}_{okl} + \nabla_l\nabla_o T_{ijk} - \nabla_k\nabla_o T_{ijl} + \nabla_o T_{ilr}\nabla_o T^r{}_{jk} - \nabla_o T_{ikr}\nabla_o T^r{}_{jl}$$

,

R being skew-symmetric with respect to the indices i and j we deduce from it :

$$(3.7)\quad H_{ijkl} + H_{jikl} + 2\,T_{ijr}\,H^r{}_{okl} + 2(\nabla_l\nabla_o T_{ijk} - \nabla_k\nabla_o T_{ijl}) = 0$$

On multiplying the two sides by g^{jl} and on changing the indices :

$$(3.8)\qquad g^{kl}\,H_{iljk} + 2\,T^k{}_{ir}\,H^r{}_{ojk} = H_{ij} + 2\,(\nabla_j\nabla_o T_i - \nabla_r\nabla_o T^r{}_{ij}).$$

We now consider the vertical 1- form on W(M) defined by its components :

(3.9) $\hat{r}_k = r_k - u_k\, r_o/F$ $\quad (r_o = r_k\, v^k)$

with

(3.10) $r_k = F^{-1}\, H_{iojk}\, X^i\, X^j,$

where X is a vector field on M. On taking into account (3.5) and (3.8) we obtain :

(3.11) $\delta^{\cdot}\hat{r} = ng\,(R(X, u), X) - \Phi(X, X),$

where $\delta^{\cdot}$denotes the co-differential of the vertical 1-form on W(M) and $\Phi(X, X)$ is defined by (3.3), M being assumed compact, by integration on W(M) we get lemma 2. From lemmas 1 and 2 we have :

Theorem. *Let (M, g) a compact Finsleriam manifold without boundary and X an infinitesimal projective transformation on M. If the quadratic form Φ(X, X) is negative definite on W(M) then the projective transformation corresponding to X is reduced to identity; if Φ(X, X) or if the flag curvature g(R(X, u)u, X) is non positive then X is an isometry and has the covariant derivation of the horizontal type zero*[3].

Proof. If $\Phi(X, X)$ is negative definite, from the lemmas 1 and 2, it follows that X is zero if $\Phi(X, X)$ or the flag curvature g(R(X, u)u, X) is non positive then

(3.12) $\nabla_{\hat{u}} Z = 0 \quad \nabla_{\hat{u}} X = F^{-2}\, g\,(\nabla_{\hat{u}} X, v)\, v,$

whence by vertical derivation

(3.13) $\nabla_j X_i + \nabla_o T_{ijh}\, X^h = F^{-2} v_i \nabla_j X_o + F^{-2}(g_{ij} - u_i u_j)\nabla_o X_o.$

On multiplying the two sides of this relation by g^{ij} and on contracting

(3.14) $\nabla_i\, X^i + \nabla_o(X_i T^i) = nF^{-2}\nabla_o X_o$

where $T^i = g^{jk} T^i_{jk}$. Now X is projective. So, by (2.11) we obtain

(3.15) $$\Psi = (1/n+1)\, \nabla_o(\nabla_i X^i + T_i \nabla_o X^i),$$

(3.16) $$\Psi = F^{-2}\, \nabla_o \nabla_o X_o\,.$$

On taking into account (3.12), (3.16) becomes

(3.17) $$\Psi = (1/n+1) \nabla_o \nabla_i X^i$$

Similarly the equation (1.11) becomes

(3.18) $$X^k R^i_{oko} = X^k H^i_{oko} = 0$$

Now the curvature tensor H is written

(3.19) $$H^i_{jkl} = (1/3)\, \delta^{\cdot}_j[\delta^{\cdot}_l H^i_{oko} - \delta^{\cdot}_k H^i_{olo}]$$

In virtue of the relations (3.18), (3.4) as well as the homogeneity of the tensor G we get

(3.20) $$H_{jl} X^j X^l = V^r \delta^{\cdot}_j H_{lr} X^j X^l = -\, \nabla_o(G^r_{jrl} X^j X^l)$$

Thus the first term in $\Phi(X, X)$ is a divergence ; we next analyze the other terms. In virtue of the proposition 2 and (3.12) we have

(3.21) $$\begin{aligned}&(\nabla_o T^r_{ij}\, \nabla_o T_r - \nabla_r\, \nabla_o\, T^r_{ij}) X^i X^j \\ &= \delta((\nabla_{\hat{v}}\, T)(X,X)) - \delta(\lambda u) - 2\Psi\, g(X, T^*)\end{aligned}$$

where T* is the torsion trace vector, $\lambda = T_{ijr}\, t^{ir}\, X^j$ and δ denotes the co-differential relative to $\rho(g)$. On the other hand by (3.14), (3.15) and (3.16) it follows

(3.22) $$\nabla_o \nabla_o g(X, T) = \frac{n-1}{n+1} \nabla_o \nabla_i X^i$$

On using (3.12) and (3.13) we obtain

(3.23)

$$(\nabla_i \nabla_o T_j - \nabla_o T_{ijh} \nabla_o T^h) X^i X^j = \frac{2}{n-1} \left\| \nabla_{\hat{v}} g(X,T) \right\|^2 + 2\Psi . g(X, T^*) + \text{Div}$$

where Div denotes the terms which are divergences ; on adding it to (3.21) we get on taking into account of (3.2) and (3.12)

$$\int_{W(M)} \left\| \nabla_{\hat{v}} g(X,T) \right\|^2 \eta(g) = 0$$

whence

$$\nabla_{\hat{v}} g(X, T^*) = 0$$

The relations (3.17) and (3.22) then imply that Ψ is identically zero. Thus, M being compact, X is an infinitesimal isometry. Hence

$$g(\nabla_{\hat{v}} X, v) = 0.$$

Then (3.12) gives $\nabla_{\hat{v}} X = 0$, and we deduce immediately that X has covariant derivation of the horizontal type zero. From the above theorem it follows in particular

Corollary. *On a Minkowskian compact manifold without boundary there do not exist projective transformations other than the isometries $P_o(M) = I_o(M)$ [3]*

4.Restricted projective vector fields [3]

Let G be the second curvature tensor of the Berwald connection D and X a projective vector field ($X \in P_o(M)$). We say X is restricted if it leaves invariant the trace tensor of G :

$$L(\hat{X}) \text{ trace } G = 0,$$

where trace G = $(G^i{}_{jik})$. After the definition of projective infinitesimal transformation it follows immediately that the function Ψ homogeneous of degree one at v is in fact a linear form at v, $\Psi = a_i(x)v^i$, and conversely. We denote by $P_o(M, r)$ the largest connected group of restricted projective transformations. We note that if $X \in P_o(M)$ and leaves the torsion trace co-vector invariant then X is restricted. Let suppose (M, g) a Landsberg manifold (P = 0), geometrically it means that if we endow the manifold V(M) with a Riemannian metric $g_{ij}(z)dx^idx^j + g_{ij}\nabla v^i\nabla v^j$ where $g_{ij}(z)$ is the Finslerian metric, then $p : V(M) \rightarrow M$ is a totally geodesic fibration, that is to say, for every $x \in M$, $p^{-1}(x)$ is a submanifold totally geodesic in V(M)[see [§7 chapter V]]. Let us suppose that it is so, then for $X \in P_o(M, r)$ we obtain

(4.1) $$\nabla_k(\nabla_j X^i + T^i{}_{jh}\nabla_o X^h) + X^h R^i{}_{jhk} = \delta^i{}_j\Psi_k + \delta^i{}_k\Psi_j + T^i{}_{jk}\Psi\ ,$$

where $\delta^i{}_j$ is the Kronecker symbol. To X we associate a 2-form with scalar values on W(M) :

(4.2) $$a(X) = (1/2)(\nabla_i X_j - \nabla_j X_i)dx^i \wedge dx^j.$$

On taking into account (4.1) we obtain :

(4.3) $$\delta(i(X)a(X)) = g(a(X), a(X)) - [2R_{ij}X^iX^j + (n-1)X^i\Psi_i]$$

where i(X) denotes the operator of interior product by X. On the other hand we have for X projective

(4.4) $$L(\hat{X})\nabla_o T^i{}_{jk} = \nabla_o L(X)T^i{}_{jk} + \Psi T^i{}_{jk} + v^i T^s{}_{jk}\Psi_s\ .$$

Now (see 5.12 Chapter III)

(4.5) $$L(\hat{X})T^i{}_{jk} = \dot{\nabla}_k(\nabla_j X^i + T^i{}_{jh}\nabla_o X^h) + X^h P^i{}_{jhk} + \nabla_o X^h\ Q^i{}_{jhk}$$

From (4.5) we obtain on contracting i and j

(4.6) $$L(\hat{X})T_k = \nabla^{\cdot}_k(\nabla_i X^i + T_i \nabla_o X^i) = \nabla^{\cdot}_k f$$
with
(4.7) $$f = \nabla_i X^i + T_i \nabla_o X^i$$

From (4.4) we then have

(4.8) $$L(\hat{X})\nabla_o T_k = \nabla_o \nabla^{\cdot}_k f + \Psi T_k$$

From (4.15), we obtain, by vertical derivation

$$(n+1)\Psi_i = \nabla_i f + \nabla_o \nabla^{\cdot}_i f$$

But by hypothesis $\nabla_o T^{\cdot i}_{jk} = 0$, whence $\nabla_o T_k = 0$, in virtue of (4.8) the above relation becomes

(4.9) $$(n+1)\Psi_i = \nabla_i f - \Psi T_i$$
whence
$$(n+1)X^i\Psi_i = X^i\nabla_i f - \Psi X^i T_i = \nabla_i(X^i f) - f\nabla_i X^i - \rho\Psi$$

with $\rho = g(X, T^*)$

The above relation can be rewritten:

(4.10) $$(n+1)X^i\Psi_i = \nabla_i(X^i f) - f(\nabla_i X^i + T_i \nabla_o X^i) + f\nabla_o\rho - \rho\Psi$$
$$= -\delta((X + \rho V)f) - f^2 - (n+2)\Psi\rho$$

where δ is the co-differential [see chapter 3]. In (4.3) appears the expression $X^i\Psi_i$ which we have just calculated; we have still to show that $\rho\Psi$ is a divergence. To this effect we put

$$Y_i = g(X, u)\Psi_i - F^{-1}g(X, u)\Psi u_i.$$

The Y_i, considered as the components of a vertical 1-form, we have

(4.11) $$-\delta^{\cdot} Y = X^i \Psi_k - nF^{-1}g(X, u)\Psi.$$

Similarly if we put

$$Z_i = F^{-1}\,[\Psi X_i - g(X, u)\Psi\, u_i]$$

we obtain

(4.12)
$$Z = n\, F^{-1} g(X, u)\,\Psi - X^i \Psi_i - 2g(X, T^*)\Psi,$$

From (4.11) and (4.12) we have

$$\delta^{\cdot}(Y - Z) = 2g\,(X, T^*)\Psi$$

Thus (4.10) becomes

$$(n+1)X^i\Psi_i = -\,\delta[(X + \rho v)f] - \frac{n+2}{2}\,\delta^{\cdot}(Y\text{-}Z) - f^2 = \mathrm{Div} - f^2.$$

Substituting this expression in (4.3) we obtain

$$\delta(i(X)a(X)) = g(a(X), a(X)) + \frac{n-1}{n+1}\,\delta(X + \rho v)^2 - 2R_{ij}X^iX^j + \mathrm{Div}$$

Let us suppose that M is compact, without boundary then on integrating on W(M) :

$$<a(X), a(X)> + \frac{n-1}{n+1}\int_{W(M)} \delta(X + \rho V)^2 \eta(g) = 2\int_{W(M)} R_{ij}X^iX^j \eta(g)$$

It thus follows that if the Ricci curvature R(X, X) is negative definite, then the infinitesimal projective transformation corresponding to X is the identity. If $R(X, X) \le 0$, then $\delta(X + \rho v) = 0$ and $a(X) = 0$. Thus X is an infinitesimal isometry, and $a(X) = 0$ implies that the covariant derivative of horizontal type of X is zero.

Theorem. *Let (M, g) be a compact Landsberg manifold without boundary $(P = 0)$ and $X \in P_0(M, r)$. If the Ricci curvature $R(X, X)$ is negative definite the corresponding projective transformation to X is identity. If $R(X, X)$ is non positive then X is an infinitesimal isometry and its covariant derivative of horizontal type is zero*[3].

5. Projective invariants [3]

Let (x^i) be the local coordinates of a point $x \in U \subset M$ and (x^i, v^i) the induced coordinates of the point $z \in p^{-1}(U)$. Let us denote by $G^i_{jk}(z)$ the coefficients of the Berwald connection ; in this chart, the curvature tensors H and G of this connection are defined by

$$H^i_{jkl} = (\delta_k G^i_{jl} - G^i_{jls} G^s_k) - (\delta_l G^i_{jk} - G^i_{jks} G^s_l) + G^i_{rk} G^r_{jl} - G^i_{lr} G^r_{jk}$$

$$G^i_{jkl} = \dot{\delta}_l G^i_{jk}, \quad G^i_{jklm} = \dot{\delta}_m G^i_{jkl}$$

with

$$2G^i = G^i_{jk} v^j v^k, \quad G^i_k = \dot{\delta}_k G^i$$

If X an infinitesimal projective transformation we have

$$L(\hat{X}) G^i = \Psi v^i, \qquad L(\hat{X}) G^i_k = \delta^i_k \Psi + v^i \Psi_k$$

(5.1)
$$L(\hat{X}) G^i_{jk} = \delta^i_j \Psi_k + \delta^i_k \Psi_j + v^i \dot{\delta}_j \Psi_k$$

(5.2)
$$L(\hat{X}) H^i_{jkl} = \delta^i_j (D_k \Psi_l - D_l \Psi_k) + \delta^i_l D_k \Psi_j - \delta^i_k D_l \Psi_j + v^i \dot{\delta}_j (D_k \Psi_l - D_l \Psi_k)$$

(5.3)
$$L(\hat{X}) G^i_{jkl} = \delta^i_j \dot{\delta}_l \Psi_k + \delta^i_k \dot{\delta}_l \Psi_j + \delta^i_l \dot{\delta}_j \Psi_k + v^i \ddot{\delta}_{jl} \Psi_k$$

and

$$\dot{\delta}_j D_l \Psi_k = D_l \dot{\delta}_j \Psi_k - \Psi_r G^r_{jkl}$$

On contacting i and k in (5.2) we obtain

$$L(\hat{X})H_{jl} = D_j\Psi_l - nD_l\Psi_j + D_o\delta^{\cdot}_j\Psi_l \qquad (H_{jl} = H^r_{jrl})$$

and

$$L(\hat{X})(v^r\delta^{\cdot}_l H_{jr}) = -(n+1)\, D_o\delta^{\cdot}_l\Psi_j$$

On eliminating Ψ_i in (5.2) and (5.3) we get
(5.4)

$$\overset{1}{W}{}^i_{jkl} = H^i_{jkl} - \frac{1}{n^2-1}[\delta^i_j(\hat{H}_{kl} - \hat{H}_{lk}) + \delta^i_k\hat{H}_{jl} - \delta^i_l\hat{H}_{jk} + v^i\delta^{\cdot}_j(\hat{H}_{kl} - \hat{H}_{lk})]$$

With

(5.5) $$\hat{H}_{jk} = n\,H_{jk} + H_{kj} + V^r\delta^{\cdot}_j H_{kr}$$

(5.6) $$\overset{2}{W}{}^i_{jkl} = G^i_{jkl} - \frac{1}{n+1}[\delta^i_j G^r_{krl} + \delta^i_k G^r_{jrl} + \delta^i_l G^r_{jrk} + v^i G^r_{jrkl}]$$

The $\overset{1}{W}$ and $\overset{2}{W}$ are homogeneous tensors of degree zero and -1 respectively, invariant by the infinitesimal projective transformation X and are called generalized projective tensors ([3]). If X is a restricted infinitesimal projective transformation the tensor $\overset{2}{W}$ reduces to G and $\overset{1}{W}$ to

(5.7)

$$\overset{*}{W}{}^i_{jkl} = H^i_{jkl} - \frac{1}{n^2-1}[\delta^i_k(nH_{jl} + H_{lj}) - \delta^i_l(nH_{jk} + H_{kj}) + (n-1)\delta^i_j(H_{kl} - H_{lk})]$$

If we multiply the two sides of (5.4) by v^l and v^j respectively we get the tensors invariant by the infinitesimal projective transformation X. Let us put

(5.8) $$\overset{1}{W}{}^i_k = \overset{1}{W}{}^i_{oko}$$

where the right hand side becomes by (5.4)

(5.8)′

$$\overset{1}{W}{}^i{}_k = H^i{}_{oko} - \frac{1}{n-1}\delta^i{}_k H_{oo} + \frac{1}{n^2-1} v^i[(2n-1)H_{ok} - (n-2)H_{ko}]$$

From (5.4), (5.8)' and (4.19) we obtain the following formulas:

$$\overset{1}{W}{}^i{}_{okl} = (1/3)(\delta^{\cdot}{}_l \overset{1}{W}{}^i{}_k - \delta^{\cdot}{}_k \overset{1}{W}{}^i{}_l) \tag{5.9}$$

$$\overset{1}{W}{}^i{}_{jkl} = (1/3)\delta^{\cdot}{}_j(\delta^{\cdot}{}_l \overset{1}{W}{}^i{}_k - \delta^{\cdot}{}_k \overset{1}{W}{}^i{}_l) \tag{5.10}$$

In other words the projective curvature tensor $\overset{1}{W}$ is expressed with the help of the tensor $\overset{1}{W}{}^i{}_k$ in the same way as the curvature tensor $H^i{}_{jkl}$ of the Berwald connection with the tensor $H^i{}_{oko}$. It thus follows the equivalence of the following three conditions :

$$\overset{1}{W}{}^i{}_k = 0 \Leftrightarrow \overset{1}{W}{}^i{}_{okl} = 0 \Leftrightarrow \overset{1}{W}{}^i{}_{jkl} = 0. \tag{5.11}$$

One can ask for the necessary and sufficient conditions for the tensors $\overset{1}{W}$ and $\overset{*}{W}$ to vanish. We consider two distinct cases and dim M>2.

1. Suppose (M, g) is isotropic :

$$H^i{}_{oko} = K(F^2\delta^i{}_k - v^i v_k) \tag{5.12}$$

where K is a function on W(M).
Whence on taking into account (3.19)

$$H^i{}_{okl} = K(\delta^i{}_k v_l - \delta^i{}_l v_k) + (1/3)\delta^{\cdot}{}_l K(F^2\delta^i{}_k - v^i v_k) - (1/3)\delta^{\cdot}{}_k K(F^2\delta^i{}_l - v^i v_l) \tag{5.13}$$

From it we deduce by contraction of i and k

$$H_{ol} = (n-1)K\, v_l + \frac{1}{3}(n-2)F^2\delta^{\cdot}{}_l K \tag{5.14}$$

Similarly by contracting i and k in (5.12) and on deriving vertically and taking into account (5.14) :

$$\text{(5.15)} \qquad H_{lo} = (n-1)K.v_l + \frac{1}{3}(2n-1)F^2\dot{\delta}_l K$$

On substituting H^i_{oko}, H_{ol} and H_{lo} in (5.8)', we note that $\overset{1}{W}{}^i_k$ vanishes. By (5.11) it follows that the projective tensor $\overset{1}{W}$ vanishes. Conversely, let us suppose that $\overset{1}{W}$ vanishes. Then by (5.8)′ we have

$$\text{(5.16)} \qquad H^i_{oko} = \frac{1}{n-1}\delta^i_k H_{oo} - \frac{1}{n^2-1} v^i [(2n-1)H_{ok} - (n-2)H_{ko}]$$

Whence on multiplying the two sides by v_i,

$$(2n-1)\, H_{ok} - (n-2)\, H_{ko} = (n+1)\, F^{-2}\, H_{oo}\, v_k$$

On substituting this expression in (5.16), we see that (M, g) is isotropic with

$$K = \frac{1}{n-1} F^{-2} H_{oo}$$

2. Let us suppose now that

$$H^i_{jkl} = K\,(\delta^i_k\, g_{jl} - \delta^i_l\, g_{jk}), \qquad (K = \text{constant})$$

On putting this expression in (5.7), we note that $\overset{*}{W} = 0$. Conversely suppose that $\overset{*}{W} = 0$. We the have the following implications:

$$\overset{*}{W}{}^i_{jkl} = 0, \Rightarrow \overset{*}{W}{}^i_{okl} = 0 \Leftrightarrow \overset{*}{W}{}^i_{oko} = \overset{1}{W}{}^i_k = 0$$

The last relation proves that (M, g) is isotropic, therefore $\overset{1}{W}$ is zero. Thus the difference $\overset{1}{W} - \overset{*}{W}$ is zero. We have therefore by (5.4) and (5.7)

$$\delta^i_l(v^r \dot{\delta}_j H_{kr}) - \delta^i_k(v^r\dot{\delta}_j H_{lr}) = (n-1)v^i\dot{\delta}_j(H_{kl} - H_{lk}) \tag{5.17}$$

Let us multiply this relation by v^k; from the fact that n > 2, we have

$$(n-2)v^i v^k \dot{\delta}_j H_{ik} = 0, \quad v^k\dot{\delta}_j H_{lk} = 0$$

Thus (5.17) gives us

$$\dot{\partial}_j H_{kl} = \dot{\partial}_j H_{lk}$$

(M, g) being isotropic, on making explicit this relation, by (5.14), we obtain

$$g_{jl}\dot{\partial}_k K + v_l\dot{\partial}_{jk}K - g_{jk}\dot{\partial}_l K - v_k\dot{\partial}_{jl} K = 0$$

let us multiply the two sides by g^{jl}; we get $(n \neq 2)$:

$$(n-2)\dot{\partial}_k K - v_k g^{jl}\dot{\partial}_{jl}K = 0 \quad \Rightarrow \dot{\partial}_k K = 0$$

Thus (M, g) has sectional curvature constant.

Theorem [3]. *Let (M, g) be Finslerian manifold (dim M > 2) ; in order that the projective tensor $\overset{1}{W}$ (respectively the restricted projective tensor $\overset{*}{W}$) is zero it is necessary and sufficient that (M, g) be isotropic (respectively has constant curvature in the Berwald connection)*

From the above theorem it follows :

Corollary [3]. - *For the restricted projective invariants* $\overset{*}{W}$ *and G (dim M > 2) to vanish, it is necessary and sufficient that (M, g) be a Riemannian manifold with constant sectional curvature.*

6. Case where Ricci directional curvature satisfies certain conditions

The Ricci directional curvature at a point $z \in V(M)$ is defined as the scalar

$$C(x, u) = F^{-2} H_{ij} v^i v^j = F^{-2} R_{ij} v^i v^j,$$

where ($R_{ij} = R^r{}_{irj}$). Now C is homogenous of degree zero in v and therefore determines a scalar function on the unitary tangent fibre bundle W(M). Let us suppose that the Ricci tensor H_{ij} satisfies

(6.1) $$D_{\hat{v}}\, H(v, v) = 0,$$

where $H(v, v) = H_{ij} v^i v^j$. If $X \in P_o(M)$, on taking the Lie derivative of the two sides of (6.1) we get

$$D_o L(X) H_{oo} - 4\, H_{oo}\, \psi = 0, \qquad H_{oo} = H(v, v),$$

Let :

(6.2) $$D_o\, D_o \psi + \frac{4}{n-1} H_{oo} \psi = 0,$$

whence

(6.3) $$F^{-4} D_{\hat{v}}\, (\psi\, D_{\hat{v}}\, \psi) - F^{-4}\, D_{\hat{v}}\, \psi . D_{\hat{v}}\, \psi + \frac{4}{n-1} C(x, u)\, (\psi/F)^2 = 0,$$

where $u = v/F$. Let us assume M is compact without boundary; then the first term of (6.3) is a divergence ; by integration on W(M) we obtain :

(6.4) $$\langle D_u(\psi/F), D_u(\psi/F)\rangle = \frac{4}{n-1} \int_{W(M)} C(x, u)\, (\psi/F)^2\, \eta(g)$$

If $C(x ; u) < 0$, then $D_o\psi = 0$ and by (7.3) we get $\psi = 0$; thus, M being compact X is an isometry. If $C(x, u) = 0$, then $D_o \psi = 0$, from the expression of ψ it follows

$$D_o D_o D_o X_o = 0,$$

M being compact we conclude then $\nabla_o\nabla_o X_o = 0 = \psi$, therefore X is an isometry.

Theorem. *If a compact Finslerian manifold without boundary satisfies the condition (6.1)($D_{\hat{v}} H(v, v) = 0$)and such that $C(x, u) \leq 0$ on the unitary fibre bundle then $P_o(M)=I_o(M)$*[3].

Let us suppose that the Ricci directional curvature C(X, u) is constant, non positive, on the unitary fibre bundle W(M); in this case it satisfies the condition (6.1) of the above theorem, we have

Corollary. *On a compact Finslerian manifold without boundary with Ricci directional curvature non positive constant, the largest connected group of projective of transformations coincides with the largest connected group of isometries.*[3]

7. The complete case.

A. Let X be a vector field leaving invariant the 1-form of splitting ∇v; we then have

(7.1) $$D_{\hat{v}} D_{\hat{v}} X + R(X, v)\, v = 0.$$

Let g be a geodesic parametrized according its arc length s. We consider along the geodesic the function

(7.2) $$\mu = X_i \frac{dx^i}{ds} = g(X, u), \quad (u = dx/ds),$$

where u is the unitary tangent vector to the geodesic considered. From the differential equation of the geodesics:

$$\frac{\nabla u^i}{ds} = \frac{du^i}{ds} + \overset{*}{\Gamma}{}^i_{jk}(x, u)u^j u^k = 0.$$

We deduce

$$\frac{d\mu}{ds} = \nabla_j X_i\, u^i u^j, \qquad \frac{d^2\mu}{ds^2} = \nabla_k \nabla_j X_i u^k u^j u^i$$

and by (7.1) we have $\frac{d^2\mu}{ds^2} = 0.$

Thus

(7.3) $$\mu = \frac{d\mu}{ds}\Big|_{s=0} s + \mu_0 = (\nabla_j X_i\, u^j u^i)_0\, s + g(X, u)_0.$$

let us suppose that (M, g) is geodesically complete then s can take every value from 0 to + ∝ if the vector X is of bounded length, so is the function μ, $\|\mu\| \leq \|X\|$, By (7.3) it follows that one has necessarily :

$$\nabla_j X_i\, u^j u^i = 0$$

X is therefore an infinitesimal isometry. We have thus established a generalization of a result of Hano[19].

Theorem. *On a geodesically complete Finslerian manifold every vector field of bounded length leaving invariant the 1-form ∇v of the splitting is an isometry*(see Chapter III) [3].

B.let us suppose the Ricci directional curvature to be constant. Then we have by vertical derivation

(7.4) $$H_{jl} + H_{lj} + v^r \dot{\partial}_j H_{lr} = 2\, C\, g_{jl},$$

where C is a constant. Let $X \in Po(M)$; on taking the Lie derivative by X of the two sides of the (7.4) we obtain

(7.5) $$(n\text{-}1)\,[D_j\,\psi_l + D_l\,\psi_j + D_o\,\dot{\partial}_j\,\psi_l] + 2\,C\,t_{jl} = 0$$

where $t = L(\hat{X})g$. On multiplying (7.5) by v^j and v^l successively

(7.6) $$\nabla_{\hat{v}}\,\psi + \frac{2C}{n-1}\,\nabla_{\hat{v}}\,g(X, v) = 0.$$

Let us suppose C = constant < 0, and put

$$K^2 = \frac{4C}{1-n},\quad \overline{\Psi} = F^{-1}\psi$$

From (7.6) we obtain by derivation

(7.7) $$\nabla_{\hat{v}}\,\nabla_{\hat{v}}\,\psi - K^2\,F^{-1}\psi = 0$$

We then conclude along the geodesic g

(7.8) $$\frac{d^2\overline{\Psi}}{ds^2} - K^2\,\overline{\Psi} = = 0.$$

Now the solution of (7.8) is of the form

(7.9) $$\overline{\Psi} = A\,\mathrm{ch}(Ks + B),$$

where A and B are constants along g. On the other hand from (7.6) we obtain

$$\frac{d\mu}{ds} = \frac{2}{K^2}\frac{d\overline{\Psi}}{ds}$$

whence

(7.10) $$\mu = \frac{2}{K^2}\,A\,\mathrm{ch}\,(Ks + B) + D\,,$$

where D is constant along g. Now from (7.2) we have $\|\mu\| \leq \|X\|$. Let us suppose X to have bounded length ; from (7.10) it follows that the length of X can be bounded only if A = 0. From (7.9) we have $\overline{\Psi} = 0$, therefore X leaves invariant the 1-form of splitting.

After the preceding theorem X is an isometry. If C= 0, by (7.6) we conclude along the geodesic g :

$$\frac{d^2\mu}{ds^2} = 2A \Rightarrow \mu = As^2 + Bs + D,$$

where A, B and D are constants on g. Similarly μ can be bounded only if A and B vanish, whence ψ is zero , and we are led to the preceding case.

Theorem.- If *(M, g) is geodesically complete Finslerian manifold with directional Ricci curvature non positive constant, then every projective vector field of bounded length is an isometry*[3].

8. Case where the Ricci directional curvature is a strictly positive constant.

Let us suppose H(u, u) = C, a positive constant; then the function Ψ satisfies (7.6), whence by derivation

$$(8.1) \qquad \nabla_{\hat{u}} \nabla_{\hat{u}} \overline{\Psi} + \frac{1}{n-1} 4C\overline{\Psi} = 0$$

where $\hat{u}$ is the horizontal vector field over u ($\nabla_{\hat{u}}$ u = 0, $\rho(\hat{u}) = u$, $\|u\| = 1$).

Let us put :

$$k^2 = \frac{1}{n-1} 4C, \quad f = \nabla_i X^i + T_i \nabla_o X^i, \quad \bar{f} = \frac{1}{n+1} f; \quad \overline{\Psi} = \nabla_{\hat{u}} \bar{f}$$

Let g be a geodesic of M ; we put along g

$$\overline{\Psi} = \frac{d\bar{f}}{ds}$$

where s is the arc length ; by (9.1) $\overline{\Psi}$ satisfies on g the differential equation of the second order :

$$\text{(8.2)} \qquad \frac{d^2\overline{\Psi}}{ds^2} + k^2\ \overline{\Psi} = 0$$

In the following we suppose that the fundamental function F is symmetric ($F(x, \lambda v) = |\lambda| F(x, v), \forall\lambda \in \mathbf{R}$) and we define a distance function on M by putting

$$d(A, B) = \inf_{c \in C(A, B)} L(c)\ , \quad L(c) = \int_a^b L(x, \frac{dx}{dt})dt$$

where C(A, B), (A, B $\in$ M) is the set of rectifiable curves joining A to B.

Let us suppose (M, g) is metrically complete; by the theorem Hopf-Rinow, valid in Finslerian geometry, [31], (M, g) is then geodesically complete. In addition, we suppose that projective infinitesimal transformation X leaves the torsion trace co-vector invariant

$$\text{(8.3)} \qquad L(\hat{X})\,T^* = 0 \Leftrightarrow \partial_i \bar{f} = 0,$$

Thus $\bar{f}$ is a function defined on M. Moreover the mean directional curvature is a positive constant ; therefore bounded from below by a positive constant ; by the theorem of Myers of Finslerian geometry [31] (in the proof of this theorem it is the expression of the Ricci directional curvature which intervenes) ; therefore M is compact. By (8.2) $\bar{f}$ becomes along the geodesic g.

$$\text{(8.4)} \qquad \bar{f} = A \cos Ks + B \sin Ks + C,$$

where A, B, and C are constants on g. The function $\bar{f}$ attains its absolute maximum and minimum on M. Let us suppose that this maximum is attained at s = 0 and the critical value of $\bar{f}$ at s = 0 is +1 ; then we have

$$(8.5) \qquad \frac{d\bar{f}}{ds} = -AK \sin Ks + B K\cos Ks$$

whence

$$(8.6) \qquad 0 = \frac{d\bar{f}}{ds}\Big|_{s=0} = BK \quad \text{and} \quad f(0) = A+C = 1.$$

Let $P_0 \in M$, a point corresponding to s = 0 ; the function $\bar{f}$ has an absolute minimum for $s = \pi/K$; we denote this point by $P_1 \in M$ and suppose that the value of the function $\bar{f}$ at this point is equal to -1 ; therefore

$$(8.7) \qquad \bar{f}\,(\pi/K) = -A + C = -1$$

From (8.4), (8.6) and (8.7) it follows :

$$(8.8) \qquad \bar{f} = \cos Ks.$$

9. The Second Variational of the Length [1d]

Let $g : [0, t] \to M$ be a Finslerian geodesic, parametrized by curvilinear abscissa, situated in a neighborhood U with local coordinates. We define on $p^{-1}(U)$ the semi-basic 1-form :

$$(9.1) \qquad \omega = \frac{\partial F}{\partial \dot{x}^i}\, dx^i = u_i\, dx^i \qquad \left(\dot{x} = \frac{dx}{ds} \right)$$

and consider the integral

(9.2) $$L(g) = \int_{o}^{t} \omega(\hat{g}) = \int_{\hat{g}} \omega,$$

where $\hat{g} : [0, t] \to W(M)$ is the canonical lift of g on the unitary tangent fibre bundle.

Let ξ be vector field on U and $\hat{\xi}$ its lift on $p^{-1}(U)$. We denote by $\gamma_\alpha = \exp(\alpha\xi)$ the local 1-parameter group generated by ξ ($\alpha \in]-\varepsilon, +\varepsilon[$) and by $\tilde{\gamma}_\alpha = \exp(\alpha\hat{\xi})$ its extension [po $\exp(\alpha\hat{\xi}) = \exp(\alpha\xi)$ op] We denote by Y the restriction of ξ to g and $g_\alpha = \gamma_\alpha . g$, $\hat{g}_\alpha = \tilde{\gamma}_\alpha . \hat{g}$. The length of the g_α is defined by

(9.3) $$L(g_\alpha) = \int_{\gamma_\alpha . \hat{g}} \omega = \int_{\hat{g}} \hat{\gamma}_\alpha^* \omega$$

whence the second variational is determined by

(9.4) $$\frac{d^2 L(g_\alpha)}{d\alpha^2}\Big|_{=0} = \int_{\hat{g}} L(\hat{\xi}) L(\hat{\xi}) \omega$$

here $L(\hat{\xi})$ denotes the Lie derivative by ξ. The right hand side of (9.4) must be evaluated along the lift of the geodesic g ($\nabla u = 0$). The expression under the sign of the integration becomes

(9.5) $$L(\hat{\xi})L(\hat{\xi})\omega = i(\hat{\xi})\, di(\hat{\xi})d\omega + di(\hat{\xi})di(\hat{\xi})\omega$$

where i( ) denotes the operator of the interior product. Now

(9.6) $$d\omega = \nabla u_i \wedge dx^i$$

and

(9.7) $$- d\nabla u_i = \nabla u_j \wedge \omega^j{}_i + u_j\, \Omega^j{}_i$$

where $(\omega^i{}_j)$ is 1-form of the Finslerian connection and Ω the corresponding curvature 2-form. We obtain

(9.8) $L(\hat{\xi})L(\hat{\xi})\omega = i(\hat{\xi})\, d(\nabla_{\hat{\xi}} u_i)dx^i + \nabla_{\hat{\xi}} u_i \nabla\xi^i - \nabla_{\hat{\xi}} \xi^i \nabla u_i$
$+ u_j\Omega^j{}_i(\hat{\xi})\xi^i + \nabla_{\hat{\xi}} u_i d\xi^i + d(u_i \nabla_{\hat{\xi}} \xi^i)$

The vector field u being unitary we have

$$i(\hat{\xi})\nabla u^i = (\delta^i{}_j - u^i u_j)\ \nabla_{\hat{u}}\ \xi^j$$

We, then, obtain along $\hat{g}$:

(9.9)
$L(\hat{\xi})L(\hat{\xi})\omega \mid \hat{g} = dg(u,\ \nabla_{\hat{Y}} Y) + [g(\nabla_{\hat{u}} Y,\ \nabla_{\hat{u}} Y) - g(R(Y, u)u, Y)$
$- \|\nabla_{\hat{u}} g(Y, u)\|^2]ds$

Theorem [3]. *If g(s) is a geodesic of the Finslerian connection parametrized according to the its arc length on [0, t] we then have along* $\hat{g}$ *:*

(9.10) $\dfrac{d^2 L}{d\alpha^2}\Big|_{\alpha=0} = g(\nabla_{\hat{Y}} Y, u)\Big|_0^t$

$$+ \int_{\grave{a}}^{t} [g(\nabla_{\hat{u}} Y,\ \nabla_{\hat{u}} Y) - g(R(Y, u)u, Y) - \|\nabla_{\hat{u}} g(Y, u)\|^2]ds.$$

10. Homeomorphie to the Sphere

Let Z be Jacobi vector field [3] along the geodesic but orthogonal to g, zero at the point s = 0and let g(t) (s= t) a fixed point on g ; we put :

(10.1) $Z(s) = \|Z(t)\| Y(s),$

where $\|Z(t)\|$ is the norm of Z at the point g(t). It is clear that Y(t) is unitary.

Let be $\sigma(\alpha)$ a geodesic passing through g(t) and tangent to Y at this point

$$(10.2) \qquad \sigma(0) = g(t), \ \frac{d\sigma}{d\alpha}\Big|_{\alpha=0} = Y(t).$$

So $\sigma(\alpha)$ determines a 1-parmater family of geodesics g_α with g_0 = g and such the Jacobi vector field is realized by g_α. There exists $\varepsilon > 0$ such that for $\alpha \in]-\varepsilon, \varepsilon[$
we have

$$(10.3) \qquad g_\alpha = \exp(\alpha Y)\, g.$$

Let $L(g_\alpha)$ be the length of the geodesic g_α (starting from the origin s= 0) with the above hypotheses the second variation of $L(g_\alpha)$ defined by (9.10) then becomes

$$(10.4) \qquad \frac{d^2 L(g_\alpha)}{d\alpha^2}\Big|_{\alpha=0} = \int_o^t [g(\nabla_{\hat{u}} Y, \nabla_{\hat{u}} Y) - g(R(Y, u)u, Y]ds$$

where $\hat{u}$ is the horizontal vector tangent to $\hat{g}$ and Y is the Jacobi field along g orthogonal to u

$$(10.5) \qquad \nabla_{\hat{u}} \nabla_{\hat{u}} Y + R(Y, u)u = 0, \qquad g(Y, u) = 0.$$

On taking into account (10.1) and (10.5) the formula (10.4) becomes:

$$(10.6) \qquad \frac{d^2 L(g_\alpha)}{d\alpha^2}\Big|_{\alpha=0} = g(Y, Y')(t) = \frac{g(Z(t), Z'(t))}{\|Z(t)\|^2}$$

We now calculate this second variational with the help of (8.4) and (8.8). In fact if $L(g_\alpha)$ is the length of the geodesic g_α between P_0 (s= 0) and $\sigma(\alpha)$ we have by (8.8)

(10.7) $$\bar{f}(\sigma(\alpha) = \cos K\, L(g_\alpha)$$

On the other hand the function $\overline{\Psi}$ deduced from $\bar{f}$ by derivation must satisfy an equation of the type (8.2) where at s we substitute the arc length α along the geodesic $\sigma(\alpha)$. We can choose $\bar{f}$ in the form :

(10.8) $$\bar{f}(\sigma(\alpha)) = A_1(t)\cos K\alpha + B_1(t)\sin K\alpha.$$

Equating (10.7) to (10.8) we have

(10.9) $$\cos KL(g_\alpha) = A_1(t)\cos K\alpha + B_1(t)\sin K\alpha\ .$$

A derivation with respect to α gives us :

(10.10) $$-K\sin KL(g_\alpha)\frac{dL}{d\alpha} = -K\, A_1(t)\sin K\alpha + B_1 K\cos K\alpha$$

and for $\alpha = 0$ we obtain $B_1 = 0$. Thus on putting $\alpha = 0$ in (10.9) we have

$$A_1(t) = \cos Kt.$$

The relation (10.10) then becomes

$$\sin KL(g_\alpha).\frac{dL}{d\alpha} = \cos Kt\, .\sin K\alpha.$$

A second derivation with respect to α then gives

$$K\cos KL(g_\alpha)\left(\frac{dL}{d\alpha}\right)^2 . + \sin KL(g_\alpha)\frac{d^2L(g_\alpha)}{d\alpha^2} = K.\cos K\alpha,$$

whence for $\alpha = 0$:

(10.11) $$\frac{d^2 L(g_\alpha)}{d\alpha^2}\Big|_{\alpha=0} = \frac{K\cos Kt}{\sin Kt} = \frac{d}{dt}\,\mathrm{Log}\,\sin Kt.$$

Equating (10.6) to (10.11) and by integration we obtain :

(10.12) $$\|Z(s)\| = 1/K\ \|Z(s)\|'_{s=0}\ .\ \sin Ks.$$

From this relation it follows that the Jacobi field along the geodesic g, vanishing at $P_0 = g(0)$, vanishes again at the point $P_1 = g(\pi/K)$. Let g_α be the 1-parameter variational of the geodesic g, passing through P_o, the points which at distance π/K from P_o, denoted $g_\alpha(\pi/K)$ describing a curve γ whose tangent vector at a point is the Jacobi vector field at this point. Now, after (10.12), the Jacobi vector field vanishes at $g_\alpha(\pi/K)$. So the curve γ becomes the identity at the point P_1. On reversing the roles P_o and P_1 we conclude that P_o and P_1 are the only critical points of f. On supposing the in addition that M is simply connected we deduce using Milnor [30] that M is homeomorphic to an n-sphere.

Theorem [3]. *Let (M, g) be complete, simply connected Finslerian manifold of dimension n with Ricci directional curvature constant strictly positive et let X be an infinitesimal projective transformation leaving the torsion co-vector invariant. Then M is compact, X defines 1-parameter global group of restricted projective transformations and M is homeomorphic to a sphere.*

CHAPTER VIII

CONFORMAL VECTOR FIELDS ON THE UNITARY TANGENT FIBRE BUNDLE

Abstract. Let X be a vector field over M and exp(tX)the local 1-parameter group generated by X and $\exp(t\hat{X})$ its lift to V(M). We call X a conformal vector field or a conformal infinitesimal transformation if there exists a function φ on M such that the Lie derivative of the metric tensor g is equal to 2 φ.g. To every field of co-vectors $Y(z) \in A_1(W)$ is associated a covariant symmetric 2-tensor $\tau(Y)$. We calculate its square by making intervene explicitly the curvature of the space. This formula helps us characterize the infinitesimal conformal transformations when M is compact without boundary. In paragraph 4 we establish a formula giving the square of the vertical part of the lift of conformal vector field on V(M) in terms of the flag curvature and we show that there exists a non-trivial conformal vector field only if the integral of the quadratic form (R(X, u)u, X) is positive. Next we take up the case when the scalar curvature $\widetilde{H} = g^{jk}\widetilde{H}_{jk}$ where $2\widetilde{H}_{jk} = \partial^2 H(v, v)/\partial v^j \partial v^k$ est a non-positive constant ($\widetilde{H}$ = constant ≤ 0.) and the torsion trace co-vector satisfies a certain condition. On making a hypothesis on the vertical Ricci curvature Q_{ij} we show that for dim M > 2, and M compact, the largest connected group $C_0(M)$ of infinitesimal conformal transformations coincides with the largest connected group of isometry $I_0(M)$, thus generalizing the Riemannian case. We deal with case where the 1-form $X = X_i(z)dx^i$ corresponding to the vector field is semi-closed and obtain in the conformal case the corresponding equations. Let σ be a semi-closed 1-form whose co-differential $\delta\sigma$ is independent of the direction and let $\overline{\Delta}\sigma$ be the horizontal Laplacian and let $\lambda(y)$, $y \in W(M)$ a the function such that $\overline{\Delta}\sigma = \lambda(y)\,\sigma$. We give an estimate of $\lambda(y)$ as a function of proper value of the flag curvature. The rest of the chapter is devoted to the lift to W(M) of the conformal vector field when its dual 1-form is semi-closed.

1. The Co-differential of a 2-form.

Let $u : M \to W(M)$ a unitary vector field and $\omega = u_i dx^i$ the corresponding 1-form. We denote by $(d\omega)^{n-1}$ the $(n-1)^{th}$ exterior power of $d\omega$, by η the volume element of $W(M)$ where

$$(1.1) \quad \eta = \frac{(-1)^N}{(n-1)!}\phi, \qquad \phi = \omega\wedge(d\omega)^{n-1}, \quad N = \frac{n(n-1)}{2}$$

We suppose M to be compact and without boundary. We denote by δ the co-differential operator, formal adjoint of d, in the global scalar product defined over W(M).
If $\pi_1 = a_i(z)\, dx^i$ and $\pi_2 = b_j \nabla u^j$ $(b_j v^j = 0)$ are respectively the horizontal 1-form and the vertical 1-form over W(M) we have (see chapter III, §7)

$$(1.2) \quad \delta\pi_1 = -(\nabla^j a_j - a_j \nabla o T^j) = -g^{ij} D_i a_j$$

$$(1.3) \quad \delta\pi_2 = -F(\nabla^{\cdot}_{j}\, b^j + b_j T^{\cdot j}) = -F g^{ij}\, \delta^{\cdot}_{i}\, b_j, \qquad b_j v^j = 0,\ \delta^{\cdot}_{i} = \frac{\delta}{\delta v^i}$$

where $(T^{\cdot i})$ is the torsion trace vector. ∇_k and $\nabla^{\cdot}_{k}$ are the components of the covariant derivations in the Finslerian connection and D_i is the covariant derivation in the Berwald connection. Let $\Lambda_{p,q}(W)$ be the space of differential forms of type (p, q) over W(M)., that is to say the forms containing p times the horizontal form dx and q times the vertical form $\beta = \nabla u$. We denote by $\Lambda_p(W)$ the space of p-forms of type (p, 0) and by H the restriction of $\Lambda_{p,q}$ to $\Lambda_p(W)$ (by abuse of language, the projection). If $\zeta(z) \in \Lambda_1(W)$, then $d\zeta$ is the sum of the forms of type (2,0) and of type (1,1) on W(M). Let us put

$$(1.4) \qquad \zeta = Y_i(z)\, dx^i$$

The restriction of the differential $d\zeta$ to $\Lambda_2(W)$ will be denoted $\bar{d}\,\zeta = H\,d\,\zeta$

(1.5) $$\alpha = \bar{d}\,\zeta = Hd\zeta = \frac{1}{2}(D_i Y_j - D_j Y_i)(z)dx^i \wedge dx^j$$

We now propose to calculate the horizontal part of the co-differential of $\bar{d}\,\zeta$. To do it, let $\pi = \zeta_i(z)dx^i \in \Lambda_1(W)$. Then the restriction of the co-differential of $\bar{d}\,\zeta$ to $\Lambda_1(W)$ will be a 1-form denoted by $\bar{\delta}\,\bar{d}\,\zeta$ such that

(1.6) $$<\bar{\delta}\,\bar{d}\,\zeta, \pi> = <\bar{d}\,\zeta, \bar{d}\,\pi>$$

We denote by (,) the local scalar product at z and by < , > the global scalar product on W(M) ([1]). Then we have

(1.7) $$\begin{aligned}(\alpha, \bar{d}\,\pi)(z) &= \frac{1}{2}\, g^{ik}g^{jl}\,(D_k\zeta_l - D_l\zeta_k)\,(D_iY_j - D_jY_i)(z)\\ &= g^{ik}g^{jl}\,D_k\zeta_l\,(D_iY_j - D_jY_i)(z)\\ &= g^{ik}\,D_k[g^{jl}\zeta_l\,(D_iY_j - D_jY_i)](z)\\ &\quad - g^{ik}\,D_k\,g^{jl}\,\zeta_l\,(D_iY_j - D_jY_i)(z)\\ &\quad - g^{ik}g^{jl}\,\zeta_l\,D_k\,(D_iY_j - D_jY_i)(z)\end{aligned}$$

Now

(1.8) $$- g^{ik}\,D_k g^{jl} = -2v^r\nabla_r T^{jli} = - 2\nabla_o T^{jli}$$

where $\nabla_o = v^r\nabla_r$. In virtue of the symmetry property of the torsion tensor T, the second term of the right hand side is zero. We therefore have

(1.9) $$(\alpha, \bar{d}\,\pi)(z) = -g^{ik}\,D_k\,(D_iY_j - D_jY_i)\zeta^j\,(z) + \text{Div on } W(M)$$

M being assumed compact without boundary, on integrating (1.9) on W(M), and taking into account (1.6) we obtain

(1.10) $$(\bar{\delta}\,\bar{d}\,\zeta)_j = (\bar{\delta}\,\alpha)_j = - g^{ik}\,D_k(D_iY_j - D_jY_i)(z)$$

We have thus an analogous expression in the Riemannian case.

2. A Lemma

To the 1-form $\zeta = Y_i(z)\, dx^i \in \Lambda_1(W)$ we associate a covariant symmetric 2-tensor τ defined by

$$(2.1) \quad \tau_{ij}(Y) = D_i Y_j + D_j Y_i + \frac{1}{2} D_o (\dot{\partial}_i Y_j + \dot{\partial}_j Y_i) + \frac{2}{n} \Psi g_{ij}$$

$\dot{\partial}_i = \frac{\partial}{\partial v^i}$, $\nabla_o = v^i \nabla_i$ and Ψ is a function, homogeneous of degree zero in v. We choose Ψ on W(M) such that the trace of τ is zero : $(g, \tau)_z = 0$. We then have

$$(2.2) \qquad \Psi = - g^{ij}[D_i Y_j + \frac{1}{2} D_o (\dot{\partial}_i Y_j)]$$

The square of the tensor τ becomes

$$(2.3) \qquad \frac{1}{2}\tau^{ij}\tau_{ij} = (\tau, \tau)_z = \tau^{ij} D_i Y_j + \frac{1}{2}\tau^{ij} D_o \dot{\partial}_i Y_j$$

Now the first term of the right hand side becomes:

(2.4)

$$\tau^{ij} D_i Y_j = g^{ik} g^{jl} \tau_{kl} D_i Y_j = g^{ik} D_i (g^{jl} \tau_{kl} Y_j) - g^{ik} D_i g^{jl} \tau_{kl} Y_j - g^{ik} g^{jl} D_i \tau_{kl} Y_j$$

Now

$$(2.5) \qquad D_i g_{kl} = - 2 D_o T_{ikl}$$

On taking into account of (1.10) and (2.5) we obtain

(2.6) $$-g^{ik} D_i \tau_{kl} = (\bar{\delta}\alpha)_l - \frac{1}{2} g^{ik} D_i D_o(\dot{\partial}_k Y_l + \dot{\partial}_l Y_k) - \frac{2}{n} D_l \psi + \frac{4}{n}\psi D_o T_l - 2 g^{ik} D_i D_l Y_k$$

where T_l is the torsion trace co-vector. On using the Ricci identity (see Chapter I) the last term becomes

$$-2 g^{ik} D_i D_l Y_k = -2g^{ik}(D_l D_i Y_k - Y_r H^r{}_{kil} - \dot{\partial}_r Y_k H^r{}_{oil})$$

(2.7). $$= -2D_l(g^{ik} D_i Y_k) + 4D_o T_l^{ik} D_i Y_k - 2[Y_r H^r{}_{kli} + \dot{\partial}_r Y_k H^r{}_{oli}]g^{ki}$$

$$-2 g^{ik} D_i D_l Y_k = 2D_l \psi + 2D_o T_l^{ik} \tau_{ik} + g^{ik} D_l D_o \dot{\partial}_i Y_k - \frac{4}{n} D_o T_l \psi - 2[Y_r H^r{}_{kli} + \dot{\partial}_r Y_k H^r{}_{oli}]g^{ik}$$

where H is the curvature of the Berwald connection. On putting (2.7) in (2.6) we get

(2.8) $$-g^{ik} D_i \tau_{kl} = P_l + 2 D_o T_l^{ik} \tau_{ik}$$

where

(2.9) $$P_l = (\bar{\delta}\alpha)_l + 2\left(1 - \frac{1}{n}\right) D_l \psi - 2(Y_r H^r{}_{kli} + \dot{\partial}_r Y_k H^r{}_{oli}) g^{ik} + g^{ik}\left[D_l D_o \dot{\partial}_i Y_k - \frac{1}{2} D_i D_o (\dot{\partial}_k Y_l + \dot{\partial}_l Y_k)\right]$$

On the other hand, the last term of the right hand side of (2.3) can be put in the form

(2.10) $$\frac{1}{2}\tau^{ij} D_o(\dot{\partial}_i Y_j) = \frac{1}{2} D_o(\tau^{ij}\dot{\partial}_i Y_j) - \frac{1}{2} D_o \tau^{ij} \dot{\partial}_i Y_j.$$

Thus taking into account (2.8), (2.9) and (2.4), the relation (2.3) becomes:

(2.11)

$$(\tau, \tau)(z) = g^{ik} D_i (\tau_{kl} Y^l) + \frac{1}{2} D_o (\tau^{kl} \dot{\partial}_k Y_l) + P_l Y^l - \frac{1}{2} D_o \tau^{kl} \dot{\partial}_k Y_l$$

We remark that the first two terms of the right hand side are divergences on W(M). We have :

Lemma 1. *For every 1-form $\zeta = Y_i(z)\, dx^i$ on W(M) we have the formula (2.11).*

3. A Characterisation of Conformal infinitesimal transformations when the manifold is compact

Let X be a vector field on M and exp(uX) the 1-parameter group of local transformations generated by X. We denote by exp($u\hat{X}$) its extension to V(M) and by L($\hat{X}$) the Lie derivative (see chap III). We say that X is an infinitesimal conformal Finslerian transformation if there exists a function φ on M such that

$$L(\hat{X})\, g_{ij} = 2\varphi g_{ij} \tag{3.1}$$

To the vector field X on M we associate by duality, defined by the metric, a 1-form on W(M) which we will denote also by $X \in \Lambda(W)$ ($X = X_i(z)\, dx^i$). On using the Berwald connection (3.1) becomes

$$D_i X_j + D_j X_i + D_o (\dot{\partial}_i X_j) - 2\varphi\, g_{ij} = 0 \tag{3.2}$$

Let us take the trace of the two sides; we get

$$\varphi = \frac{1}{n} [g^{ij} D_i X_j + D_o (T^i X_i)] = -\frac{1}{n} \delta(X + \rho v), \qquad \rho = (X, T^*) \tag{3.3}$$

where T* is the torsion trace vector. We are led to associate to X a covariant symmetric 2-tensor $\tau(X)$ defined by

(3.4) $$\tau(X)_{ij} = D_iX_j + D_jX_i + D_o(\partial_i X_j) + \frac{2}{n}\delta(X+\rho v)g_{ij}$$

where $\delta(X+\rho v)$ is independent of the directions :

(3.5) $$\partial_i\,\delta(X+\rho v) = 0 \qquad \left(\partial_i = \frac{\partial}{\partial v^i}\right)$$

Thus, for X to be an infinitesimal conformal transformation it is necessary and sufficient that $\tau(X) = 0$ and (3.5) holds.

After (2.11) the square of $\tau(X)$ becomes

(3.6) $(\tau(X)\,\tau(X))\,(z) = (\zeta, X) + \text{Div dependent on } \tau$ on W(M)
where

(3.7) $$\text{Div dependent on } \tau = g^{ik} D_i(\tau_{kl} X^l) + \frac{1}{2} D_o(\tau^{kl}\,\partial_k X_l)$$

In virtue of (2.8) and (2.11) we have

(3.8) $$\zeta_l = P_l - T_l^{ik} D_o\tau_{ik} = -[g^{ik} D_i\tau_{kl} + D_o(T_l^{ik}\,\tau_{ik}) + D_o(T_l^{ik})\tau_{ik}]$$

Once again let, taking into account (2.9)
(3.9)

$$\zeta_l = (\bar{\delta}\,\alpha)_l + 2\left(1-\frac{1}{n}\right)D_l\delta(X+\rho v) + g^{ik}(D_l D_o\,\partial_i X_k - D_i D_o\,\partial_k X_l)$$
$$- 2[X_r H^r{}_{kli} + \partial_r X_k H^r{}_{oli}]g^{ik} - T_l^{ik} D_o\tau_{ik}$$

with $$\alpha = \frac{1}{2}(D_iX_j - D_j X_i)\, dx^i \wedge dx^j$$

Now let us suppose M compact without boundary. On integrating (3.6) on W(M) we obtain :

(3.10) $$\langle\tau(X), \tau(X)\rangle = \langle\zeta, X\rangle$$

Let us suppose that $\tau = 0$; from (3.8) it follows that $\zeta = 0$; conversely suppose that $\zeta = 0$; after (3.10) we have $\tau = 0$. Therefore X is an infinitesimal conformal transformation.

Theorem. *In order that a vector field X on a compact Finslerian manifold without boundary defines an infinitesimal conformal transformation it is necessary and sufficient that $\zeta = 0$ and $\delta(X+\rho v)$ be independent of the direction :*

$$(\bar{\delta}\alpha)_l + 2(1 - \frac{1}{n})D_l\,\delta(X + \rho v) + g^{ik}(D_l D_o \dot{\partial}_i X_k - D_i D_o \dot{\partial}_k X_l)$$

$$= 2\,\hat{H}_{jl}X^j + T_l^{\;ik}\,D_o\,\tau_{ik} \tag{3.11}$$

and $$\dot{\partial}_l\,\delta(X + \rho v) = 0$$

where $\hat{H}_{jl}$ is defined by

$$\hat{H}_{jl} = g^{ki}\,H_{jkli} + 2\,T^i_{\;jr}\,H^r_{\;oli} \tag{3.12}$$

In case X is an infinitesimal isometry we have $\zeta = 0$, and $\delta(X + \rho v) = 0$.

Corollary. *In order that a vector field X on a compact Finslerian manifold without boundary defines an infinitesimal isometry it is necessary and sufficient that $\zeta = 0$, and $\delta(X + \rho v) = 0$.*

4. Curvature and Infinitesimal Conformal Transformations in the compact case

A. Lemma 2. *Let X be a vector field on M, we then have*

$$\hat{H}(X, X) - n(H(X, u)u, X) = \text{Div on M} \tag{4.1}$$

For the proof (see chapter VII §4, lemma 2, relation (4.2), (4.3) and (4.8))

Lemma 3. Let X be a vector field on M. We have

(4.2)
$$F^{-2}(X_i H^i_{jok} g^{jk} X_o - H^{r\ j}_{\ o\ o}\ \partial^\bullet_i X_j X_o) = (H(X,u)u, X)) + \text{Div on } W(M)$$

where $X_o = (X, v)$.

Proof. In virtue of Bianchi identity we have

$$F^{-2}\ \partial^\bullet_j\ X_i H^{i\ j}_{\ o\ o} X_o$$
$$= F\, g^{jk}\ \partial^\bullet_j (X_i H^i_{oko} X_o F^{-3}) + F^{-2} (X_i H^i_{jok}\ g^{ik} X_o) - (H(X, u)u, X))$$

Now the first term of the right hand side is a divergence. Hence the lemma

Lemma 4 *Let X be a conformal vector field on M and let* $\psi=\delta(X+\rho v)$ *be a function defined on M where* $\rho = (X, T^*)$, *we have*

$$(X, T^*)\, D_o\delta(X + \rho v) = \text{Div on } W(M) \tag{4.3}$$

where T* is the torsion trace vector.
Proof. Let $\hat{Y}$ be a vertical 1-form defined by its components:

$$\hat{Y}_i = F^{-1} X_i D_o\,\psi - u_i (X, v)\, D_o\,\psi\, F^{-2} \qquad D_o\,\psi = v^i\, D_i\,\psi \tag{4.4}$$

$$F g^{ij}\ \partial^\bullet_j\ \hat{Y}_i = 2(X, T^*)\, D_o\psi\ + (X, d\psi) - n(X, v) D_o\,\psi F^{-2} \tag{4.5}$$

where $(X, d\psi) = X^i D_i \psi$. We now calculate the last two terms of the right hand side. To this effect, X being conformal, we have

$$D_i X_j + D_j X_i + D_o\ \partial_i X_j + \frac{2}{n}\psi g_{ij} = 0 \tag{4.6}$$

On multiplying the two sides by v^i and v^j successively we get

$$\psi = - nD_o (X, v)F^{-2}$$

Thus the last term of the right hand side of (4.8) becomes:

$$(4.7) \quad -n(X, v) D_o \psi . F^{-2} = -nD_o (F^{-2} (X, v)\psi) + n F^{-2} D_o (X, v) \psi$$
$$= \text{Div on } W(M) - \psi^2$$

It remains to calculate $(X, d\psi)$. We have ψ independent of the direction:

$$(4.8) \quad (X, d\psi) = X^i D_i \psi = g^{ij} D_i (X_j . \psi) - g^{ij} D_i X_j \psi$$
$$= \text{Div on } W(M) + \psi^2 + D_o (X, T^*)\psi$$
$$= \text{Div on } W(M) + D_o [(X, T^*)\psi] - (X, T^*) D_o\psi + \psi^2$$
$$= \text{Div on } W(M) - (X,T^*) D_o\psi + \psi^2$$

Taking into account (4.7) and (4.8) the relation (4.5) becomes :

$$F g^{ij} \dot{\partial}_j \hat{Y}_i = (X, T^*) D_o\psi + \text{Div on } W(M)$$

Now, after (1.3), the left hand side is a divergence on W(M); hence the lemma.

Remark. To the vector field X on M we associate the 1-form $\hat{X}$ on W(M) defined by

$$(4.9) \qquad \hat{X} = X_i (z) dx^i + F^{-1} \dot{X}_i Dv^i$$

where

$$(4.10) \qquad \dot{X}_i = F^{-1} (D_o X_i - v_i D_o (X, v) F^{-2})$$

After (1.2) and (1.3), the co differential of $\hat{X}$ is

$$-\delta \hat{X} = g^{ij} D_i X_j + F g^{ij} \dot{\partial}_j \dot{X}_i$$

On making explicit the last term, after (4.10) we obtain :

$$(4.11)\quad -\delta \hat{X} = 2\,[g^{ij} D_i X_j + D_o (X, T^*) - \frac{n}{2}\frac{D_o(X,v)}{F^2}]$$

Let us suppose now that X is an infinitesimal conformal transformation. We have

$$\psi = \delta(X + \rho v) = -\, n\, \frac{D_o(X,v)}{F^2}$$

Thus

$$\delta\, \hat{X} = \delta(X + \rho v) = \psi$$

B. Let us suppose that X is conformal. Therefore it satisfies (3.11). Let us multiply the two sides of this relation by v^l. Then we get:

(4.12)

$$-g^{ik} D_i (D_k X_o - D_o X_k) + 2\,(1 - 1/n)\, D_o\, \delta \hat{X} + 2\, D_o D_o\, (X, T^*) = 2\, \hat{H}_{jo} X^j$$

On the other hand from (4.9) it follows, on multiplying the two sides by v^i and on exchanging j and k

$$(4.13)\qquad (D_k X_o + D_o X_k) + \frac{2}{n}\delta\, \hat{X}\, v_k = 0$$

On multiplying the two sides of (4.12) by $F^{-2}(X, v)$ and on using (4.13) we obtain:

$$(4.14)\quad -F^{-2}(X, v)\, g^{ik} D_i D_k (X, v) + (1 - \frac{2}{n})(X, v)\, F^{-2} D_o \delta\, \hat{X}$$
$$+ F^{-2}(X, v)\, D_o D_o (X, T^*) = F^{-2}(X, v)\, \hat{H}_{jo} X^j$$

We have successively

$$(4.15)\quad -F^{-2}(X, v)\, g^{ik} D_iD_k(X,v) = -F^{-2} g^{ik} D_i [D_k(X, v)(X, v)] + F^{-2} g^{ik} D_i (X, v) D_k (X, v) = \text{Div on } W(M) + F^{-2} g^{ik} D_i (X, v) D_k (X, v)]$$

$$(4.16)\quad (1\text{-}2/n)\, F^{-2}(X, v) D_o \delta \hat{X} = (1\text{-}2/n) D_o[(X, v)\delta \hat{X}\, F^{-2}] + \frac{1}{n} 1/(1\text{-}2/n)\, (\delta \hat{X}, \delta \hat{X})$$

and, after lemma 4,

(4.17)

$$F^{-2}(X, v) D_o D_o(X, T^*) = D_o[F^{-2}(X, v)\, D_o(X, T^*)] + D_o(X, T^*) \frac{1}{n} \delta \hat{X} = \text{Div on } W(M) - \frac{1}{n} (X, T^*) D_o \delta \hat{X} = \text{Div on } W(M).$$

Thus on putting (4.15) (4.16) and (4.17) in (4.14) and on using lemma 3 we obtain:

$$(4.18)\quad (D_{\hat{u}} X, D_{\hat{u}} X) + \frac{1}{n}(1-\frac{2}{n})(\delta \hat{X}, \delta \hat{X}) = (H(X, u)u, X) + \text{Div on } W(M).$$

where $\hat{u}$ is a horizontal vector field over u, ($\rho \hat{u} = u$), M is compact and without boundary; on integrating the above relation on W(M) we obtain :

$$(4.19)\quad < D_{\hat{u}} X, D_{\hat{u}} X> + \frac{1}{n}(1 - \frac{2}{n}) <\delta \hat{X}, \delta \hat{X}> = <H(X, u)u, X>$$

where <, > denotes the global scalar product over W(M).
If $<H(X, u)u, X> \leq 0$, then $\delta \hat{X} = 0$, and $D_{\hat{u}} X = 0$; by vertical derivation we obtain

$$D_j X_i + D_o(\partial_j^{\cdot} X_i) = 0.$$

On the other hand, X being conformal and $\delta \hat{X} = 0$, it then follows

$$D_i X_j = 0 = \nabla_i X_j \;, D_o(\partial_i^{\cdot} X_j) = 0$$

Theorem. *Let (M, g) be an n-dimensional compact Finslerian manifold without boundary, and X an infinitesimal conformal transformation. If the sectional curvature (H(X, u)u, X) is non-positive everywhere, then X is an infinitesimal isometry and has covariant derivation of horizontal type zero in the Finslerian and in the Berwald connection[2].*

5. Case when M compact with scalar curvature $\widetilde{H}$ constant

Let X a conformal vector field on M ; it satisfies (3.1). The Lie derivation by X of the coefficients of the Finslerian connection becomes (see [1])

$$(5.1)\qquad L(\hat{X})\overset{*}{\Gamma}{}^{i}_{jk} = \delta^{i}_{j}\,\varphi_k + \delta^{i}_{k}\,\varphi_j - g_{jk}\,\varphi^i$$
$$- [T^{i}_{jr} L(\hat{X})\overset{*}{\Gamma}{}^{r}_{ok} + T^{i}_{kr} L(\hat{X})\overset{*}{\Gamma}{}^{r}_{oj} - T_{jkr}\, g^{ih}\, L(\hat{X})\overset{*}{\Gamma}{}^{r}_{oh}]$$

On multiplying (5.1) successively by v^j and v^k we get :

$$(5.2)\quad L(\hat{X})\,\overset{*}{\Gamma}{}^{i}_{ok} = L(\hat{X})\,G^{i}_{k} = \delta^{i}_{k}\varphi_o + v^i\varphi_k - v_k\varphi^i + F^2 T^{i}_{kr}\varphi^r$$

$$(5.3)\qquad L(\hat{X})\,\overset{*}{\Gamma}{}^{i}_{oo} = 2L(\hat{X})\,G^i = 2v^i\,\varphi_o - F^2\varphi^i$$

From (5.2) we obtain by vertical derivation

$L(\hat{X})G^i_{jk}=\delta^i_j\,\varphi_k+\delta^i_k\varphi_j-g_{jk}\,\varphi^i-v_k\,\partial_j\,\varphi^i+2v_j\,T^{ir}_{\ \ k}\,\varphi_r+F^2\,\partial_j\,T^{ir}_{\ \ k}\,\varphi_r$

(5.4)

Now the Lie derivative of curvature tensor of the Berwald connection becomes

(5.5)

$L(\hat{X})H^{\ i}_{jkl}=D_kL(\hat{X})G^i_{jl}-D_lL(\hat{X})G^{\ i}_{jk}+G^{\ i}_{jkr}L(\hat{X})G^r_l-G^{\ i}_{jlr}L(\hat{X})G^r_k$

On contracting i and k and on multiplying the two sides v^j and v^l we obtain

(5.6) $$L(\hat{X})\,H(v,v)=2\,D_i\,L(\hat{X})\,G^i-D_o\,L(\hat{X})\,G^i_i$$

where we have put $H(v, v) = H_{ij}\,v^iv^j$.

On using (5.2) and (5.3) the above relation becomes

(5.7) $$L(\hat{X})\,H(v,v)=(2\text{-}n)D_o\varphi_o\ -F^2\,[D_i\varphi^i+D_o(T^i\,\varphi_i]$$

where o denotes the multiplication contracted by v. Let us put

(5.8) $$\widetilde{H}_{jk}=\frac{1}{2}\frac{\partial^2 H(v,v)}{\partial v^j\partial v^k}$$

$\widetilde{H}_{jk}$ is a symmetric tensor of order 2, homogeneous of degree zero in v. We put

(5.9) $$\rho=D_i\,\varphi^i+D_o\,(T^i\varphi_i)=g^{ij}D_i\varphi_j+2\,\varphi_iD_o\,T^i+D_o\,(T^i\varphi_i)$$

On taking into account the identity (4.2) of the relation (5.7) we obtain by vertical derivation

(5.10) $$L(\hat{X})(H_{ko}+H_{ok})=2(2\text{-}n)D_o\,\varphi_k-2v_k\,\rho-F^2\rho_k,\quad (\rho_k=\partial_k\,\rho)$$

(5.11) $L(\hat{X})\widetilde{H}_{jk} = (2-n)\, D_j\varphi_k - \rho g_{jk} - v_k\, \rho_j - v_j\, \rho_k - \frac{1}{2} F^2\, \partial_{\dot{j}\,\dot{k}}\, \rho$

whence, taking into account (5.9) :

(5.12)

$$g^{jk}L(\hat{X})\widetilde{H}_{jk} = 2(1-n)g^{jk}\, D_j\varphi_k - 2n\varphi_i D_o T^i - nD_o\,(T^i\varphi_i) - \frac{1}{2} F^2\, g^{jk}\,\partial_{\dot{j}\,\dot{k}}\,\rho$$

Let us suppose now that the scalar curvature is a non positive constant

(5.13) $$\widetilde{H} = g^{jk}\,\widetilde{H}_{jk} = \text{constant} \leq 0$$

By Lie derivation of the two sides of (5.13) and, on taking into account (5.12), we get

(5.14) $\Delta\varphi$

$$= \frac{1}{n-1}\,\widetilde{H}\,\varphi + \frac{n}{n-1}\varphi_i\, D_o T^i + \frac{1}{2}\frac{n}{n-1} D_o\,(T_i\varphi^i) + \frac{1}{4}\frac{1}{n-1} F^2 g^{jk}\,\partial_{\dot{j}\,\dot{k}}\,\rho$$

where $\Delta\varphi$ is the Laplacian on W(M) defined by (3.3), φ being independent of the direction , from (5.14) we obtain on multiplying the two sides by φ,

(5.15) $$(\Delta\varphi, \varphi) = \frac{1}{n-1}\,\widetilde{H}\,\varphi^2 + \frac{n}{n-1}\varphi\varphi_i\, D_o T^i + \frac{1}{2}\frac{n}{n-1} D_o(T^i\varphi_i\varphi) - \frac{1}{2}\frac{n}{n-1} T^i\varphi_i\varphi_o + \frac{1}{4}\frac{1}{n-1} F^2 g^{jk}\,\partial_{\dot{j}\,\dot{k}}(\rho\varphi)$$

We are thus led to calculate the terms $T^i\,\varphi_i\,\varphi_o$ and $\varphi\,\varphi_i\, D_o T^i$

Lemma 5. *Let Y and Z are two vertical 1-forms defined by*

(5.16) $Y_k = \frac{1}{2} F\, T_{ijk}\,\varphi^i\varphi^j, \quad Z_k = \frac{1}{2} FT^j\,\varphi_j\varphi_k - \frac{1}{2} u_k\, T^j\varphi_j\,\varphi_o$

When u = v/F, we have the formula

$$\delta Z - \delta Y + F^2 Q_{ij}\, \varphi^i \varphi^j = \frac{n-2}{2}\, T^j \varphi_j\, \varphi_o \tag{5.17}$$

where Q_{ij} is the Ricci tensor corresponding to the third curvature tensor of the Finslerian connection :

$$Q_{ij} = T^s_{ri} T^r_{sj} - T_r T^r_{ij} \tag{5.18}$$

Proof. Consider the expression

$$\begin{aligned} - F^2\, T^i_{lr} T^r_{ij}\, \varphi^j \varphi^l &= \frac{1}{2} F^2\, \partial_{\dot{r}}\, \varphi^i\, T^{rj}_{i} \varphi_j \\ &= - \delta Y - \frac{1}{2} F^2\, g^{rk}\, \partial_{\dot{r}}\, (T^j_{ik})\, \varphi_j \varphi^i \end{aligned} \tag{5.19}$$

Now
$$\begin{aligned} g^{rk} \partial_{\dot{r}}\, (T^j_{ik}) &= \partial_{\dot{r}}\, (T^{jr}_{i}) - \partial_{\dot{r}}\, g^{rk}\, T^j_{ik} \\ &= \partial_{\dot{i}}\, T^j + 2\, T^r T^j_{ir} \end{aligned}$$

Thus on taking into account the expression of Q_{ij} , (5.18) becomes

$$-\delta Y + F^2\, Q_{ij}\, \varphi^i \varphi^j = \frac{1}{2} g^{ik}\, \partial_{\dot{i}}\, (F T^j\, \varphi_j \varphi_k) - \frac{1}{2} T^j \varphi_j \varphi_o \tag{5.20}$$

On using the co-differential of Z according to the formula (1.3) and on putting it in the above relation we find the lemma.

Lemma 6. Let (M, g) a Finslerian manifold of dimension n such that

$$\tau = g^{ij}(D_i D_o T_j + \partial_{\dot{i}}\, D_o D_o T_j) = 0 \tag{5.21}$$

Then $\varphi \varphi_i D_o T^i$ is a divergence on W(M).

Proof. $\varphi\varphi_i D_o T^i = \frac{1}{2} D_i\, \varphi^2 D_o T^i = \frac{1}{2} g^{ij} D_i (\varphi^2 D_o T_j) - \frac{1}{2} \varphi^2 g^{ij} D_i D_o T_j$

In virtue of (5.21) we have

$$\varphi\varphi_i D_o T^i = \frac{1}{2} g^{ij} D_i(\varphi^2 D_o T_j) + \frac{1}{2} g^{ij} \dot{\partial}_j (\varphi^2 D_o D_o T_i)$$

After (1.2) and (1.3) each term of the right hand side is a divergence on W(M).

Let us suppose that M is compact without boundary. On integrating (5.15) on W(M) on using the preceding lemmas we obtain

$$\text{(5.22)} \qquad \int_{W(M)} q_{ij}\, \varphi^i \varphi^j \eta = \frac{\widetilde{H}}{n-1} \int_{W(M)} \varphi^2 \eta$$

where we have put

$$\text{(5.23)} \qquad q_{ij} = g_{ij} + \frac{n}{(n-1)(n-2)} F^2 Q_{ij}$$

Let us suppose that q_{ij} is positive definite. If $\widetilde{H}$ is constant , non positive from (5.22) it follows that $\varphi = 0$ for $\widetilde{H}$ = constant < 0. This is clear and for $\widetilde{H} = 0$, $\varphi_i = 0$, so φ is constant. Now from (3.3) φ is a divergence and M is compact, hence $\varphi = 0$, and we have

Theorem. *Let (M, g) be a compact Finselerian manifold of dimension n > 2 without boundary. Let us suppose that the scalar curvature $\widetilde{H}$ defined by (5.13) be a non-positive constant and τ defined by (5.21) vanishes everywhere. If the quadratic form q_{ij} defined by (5.23) is positive definite everywhere on W(M) then the largest connected group of infinitesimal conformal transformations $C_o(M)$ coincides with the largest connected group of isometries $I_o(M)$*

6. Case when $X = X_i(z)\, dx^i$ is semi-closed.

A. Let X be a vector field on M and let $X = X_i(z)\, dx^i$ be the corresponding dual 1-form. Its differential is

$$(6.1)\quad dX = \frac{1}{2}(D_iX_j - D_jX_i)\, dx^i \wedge dx^i - \dot{\partial}_j X_i dx^i \wedge \nabla v^j$$

we say that the 1-form X is <u>semi-closed</u> if the horizontal part of dX vanishes:

$$(6.2)\qquad D_iX_j = D_jX_i$$

On multiplying the two sides by v^i:

$$(6.3)\qquad D_oX_j = D_jX_o$$

Whence by vertical derivation:

$$(6.4)\qquad D_iX_j + D_o\, \dot{\partial}_i\, X_j = D_j\, Xi$$

On taking into account (6.2) we have

$$(6.5)\qquad D_o\, \dot{\partial}_i\, X_j = 0$$

Multiplying the two sides by g^{ij} and contracting

$$(6.6)\qquad D_0\,(X, T^*) = 0\,, \qquad (X, T^*) = (X^iT_i)$$

Let us put

$$(6.7)\qquad Z_l = 2(1-\frac{1}{n})\, D_l\, \delta X\; - 2\, \hat{H}_{jl}\, X^j \qquad \text{with } \dot{\partial}_j \delta X = 0$$

where $\hat{H}$ is defined by (3.12). If X is semi-closed, the expression ζ, on taking into account (6.5) and (6.6) reduces to

$$(6.8)\qquad \zeta_l = Z_l - T_l^{jk}\, D_o\, \tau_{ik}$$

whence from (36) and (3.9)

(6.9) $(\tau(X), \tau(X)) = (\zeta, X) +$ Div dependent on τ on W(M)

(6.10) $= (Z, X) +$ Div dependent on τ on W(M)

Now by the above theorem, infinitesimal conformal transformations are characterized by $\zeta = 0$. and so by (6.8) , $Z = 0$. Conversely if $Z = 0$ from (6.9) by integration on W(M) we have $\tau = 0$. Thus

Corollary. *On a Finslerian compact manifold without boundary in order that a vector field whose dual 1-form is semi-closed defines an infinitesimal conformal transformation it is necessary and sufficient that one has*

(6.10) $$(1-\frac{1}{n})\, D_l\delta X = \hat{H}_{jl}\, X^j \qquad \partial^{\bullet}_j \delta X = 0$$

Let us rewrite the formula (6.9) by letting flag curvature intervene directly. Using the lemma 2

(6.11) $$\left(\frac{1}{n}\left(1-\frac{1}{n}\right) \bar{\Delta} X - R(X, u)u, X\right) - \frac{1}{2}\,\frac{1}{n}\,(\tau(X), \tau(X)) = \text{Div on } W(M)$$

where $\bar{\Delta} X$ is the horizontal Laplacian $\bar{d}\,\delta\, X$. Let us take a frame at a point $x = py \in M$ such that $u = e_n$ and let us put $R_{\alpha\beta} = R_{n\alpha n\beta}$ where $R_{\alpha\beta}$ is symmetric and suppose that $R_{\alpha\beta}\, X^\alpha X^\beta$ is defined everywhere on W(M) (of rank (n-1)). Then on a Finslerian compact manifold without boundary there exist non-vanishing semi-closed infinitesimal conformal transformations only if $R_{\alpha\beta} X^\alpha X^\beta$ is positive definite ($\alpha, \beta = 1, \ldots n-1$). At the point $y \in$ W(M) let $\lambda_1(y)$ is the least proper value of the operator $R_{\alpha\beta}$ and put $\lambda_1 = \min \lambda_1(y)$ for $y \in$ W(M). Let $\lambda\,(y)$ be a function such that $\bar{\Delta}\,\sigma = \bar{d}\,\delta\sigma = \lambda(y)\sigma$ where σ is a semi-closed 1-form corresponding to a vector field on M and $\bar{\Delta}$ the horizontal Laplacian. Then the formula (6.11) becomes

(6.9)
$$\left(\frac{1}{n}\left(1-\frac{1}{n}\right)\lambda(y)\,\sigma - R(\sigma\, u)u, \sigma\right) - \frac{1}{2}\frac{1}{n}\left(\tau(\sigma), \tau(\sigma)\right)$$
$$= \text{Div on } W(M)$$

if
$$\frac{1}{n}\left(1-\frac{1}{n}\right)\lambda(y) < \lambda_1$$

Then by (6.12) we obtain by integration on W(M), $\tau(\sigma) = 0$. and if
$$\lambda(y) = n\lambda_1\left(1 - \frac{1}{n}\right)^{-1}$$

then σ defines a semi-closed conformal infinitesimal transformation. Theorem. On a compact Finslerian manifold without boundary with $R_{\alpha\beta}\, X^{\alpha}X^{\beta}$ positive definite the functions $\lambda(y)$ such that $\bar{\Delta}\,\sigma=\lambda(y)\sigma$ where $\bar{\Delta}$ is the horizontal Laplacian and σ a semi-closed 1-form corresponding to the vector field X on M is greater_than $n\lambda_1\left(1 - \frac{1}{n}\right)^{-1}$.

A. Let X be a vector field on M and $\tilde{X}$ its lift to W(M) We suppose that the 1-form corresponding to X is semi-closed and X is conformal. By (3.2) and (6.2) it follows that we have

(6.14)1.1.1.1. $D_o X_i = v_i\, \varphi$

(6.15) $D_o (X, T^*) = 0$

where T* is the torsion trace vector. From (6.13) we have

(6.15) $F^2\, \varphi = D_o X_o$

Thus from (6 .13) it follows that the lift of X on W(M) is horizontal and satisfies (6.14). Conversely from (6.13) and (6.15) we obtain, by vertical derivation,

(6.16) $F^2\, \varphi_{ij} + v_i\varphi_j + v_j\varphi_i + D_o\, \partial^{\bullet}_i X_j = 0,\ \ \varphi_i = \dfrac{\partial \varphi}{\partial v^i},\ \ \varphi_{ij} = \dfrac{\partial^2 \varphi}{\partial v^i \partial v^j}$

Let us multiply both the sides of the above relation by g^{ij} and contract. Now φ is homogeneous of degree zero in v. So taking into account (6.14) we obtain

$$F^2\, g^{ij} \frac{\partial^2 \varphi}{\partial v^i \partial v^j} = 0$$

By a reasoning identical to (see chapter VI §2) we show that φ is independent of the direction. So from (6.15) and (6.13) we obtain

$$D_j X_i = g_{ij}\, \varphi = D_i X_j$$

X semi-closed and conformal:

Proposition. *Let(M, g) be a Finslerian manifold. If the lift of X on W(M) is horizontal and satisfies*

$$\nabla_o(X, T^*) = 0$$

Then X is conformal and its dual with respect the Finslerian metric is semi-closed and conversely if X is conformal and semi-closed then the lift of X on W(M) is horizontal and satisfies (6.14)

Corollary. *Let X be a vector field on a Riemannian manifold. If the lift of X on the sphere bundle S(M) is horizontal then the 1-form dual to X is closed and conformal. Conversely if the lift on the sphere bundle of M of a conformal vector field whose dual 1-form is closed is horizontal.*

References

[1] H.Akbar-Zadeh. Les Espaces de Finsler et certaines de leurs généralisations.Ann.Ec.Norm.Sup. 3e Série 80. (1963) 1-79

[1a] H. Akbar-Zadeh. Sur les isométries infinitésimales d'une variété finslérienne compacte. C.R. Acad.Sci. Paris t. 278 (18 mars 1974) Série A –871

[1b] H. Akbar-Zadeh. Remarques sur les isométries infinitésimales d'une variété finslérienne compacte. C.R. Acad. Sci. Paris t. 284 (2& février 1977) Série A. 451

[1c] H. Akbar-Zadeh, Sur une Connexion euclidienne d'espace d'éléments linéaires . C.R Acad.Sc. Paris t.245 (1957) Série A pp 26-28

[1d] H ; Akbar-Zadeh, Sur quelques théorèmes issus du calcul des variations . C.R. Acad. Sc. Paris t.264, Série A (1967) pp 517-519

[2] H.Akbar-Zadeh. Transformations infinitésimales conformes des variétés finsleriennes compactes. Ann.Polon.Math XXXVI (1979) 213-229

[3] H.Akbar-Zadeh. Champ de vecteurs projectifs sur le fibré unitaire. J Math pures et appl. 65 (1986) p.47-79.

[4] H.Akbar-Zadeh, Sur les espaces de Finsler à courbures sectionelles constantes, Acad Royale Belgique Bull. de.Sci. 5eme série LXXIV (1988-10) 281-322

[5] H.Akbar-Zadeh, Generalized Einstein Manifolds, Journal of Geometry and Physics, 17 (1995) 342-380

[6] H.Akbar-Zadeh, Geometry of Generalized Einstein Manifolds, C.R. Acad. Sci. Paris, Série I 339 (2004)

[7] H.Akbar-Zadeh, Sur les sous-variétés des variétés finslériennes. C.R. Acad. Sci. Paris t. 266 (1968) pp 146-148

[8] L.Berwald. Parallelübertragung. In allgemeinen. Räumen (Atti congresso intern.Matem. Bologna 1928 IV p. 263-270).

[9] N. Bourbaki. Topologie générale Livre III Hermann Paris 1955.

[10] F. Brickell A Theorem on Homogeneous Functions. J London Math Society 42. (1967) pp.325-329.

[11] H. Busemann On Normal Coordinates in Finsler Spaces Math Annalen 129 (1955)
[12] E. Cartan. Sur les espaces de Finsler, C.R.Acad. Sc. t.196, (1933) p.582-586
[13] E. Cartan. Les espaces de Finsler. Paris, Hermann. (1934)
[14] E. Cartan. Œuvres Complètes. Partie III vol 2. Géometrie Différentielle, Gauthier –Villars (1955) p. 1393
[15] A. Diecke, Über die Finsler-Räume mit $A_i = 0$ Arch. Math 51953) p.45-51
[16] C. Ehresmann, les Connexions infinitesimales dans un espace fibré differentiable, Colloque de Topologie, Bruxelles 1950, pp 29-55.
[17] P. Finsler, Uber Kurven und Flächen in allgemeinen Räumen, Dissertation Göttingen 1918
[18] M. Haimovici, Variétés totalement extremales et variétés totalement géodésiques dans les espaces de Finsler, Ann. Sci. Univ. Jassy I –25 (1939) pp.559-644
[19] J. Hano, On Affine transformations of a Riemannian manifold, Nagoya Math Journal vol 9 pp 98-109.
[20] D.Hilbert, Die Grundlagen der Physik, Nachr.Akad.Wiss. Göttingen, (1915), 395-407
[21] A. Kawaguchi. On the theory of non-linear connections, II Tensor N.S. 6 (1956) pp 165-199
[22] H. Kawaguchi. On the Finsler spaces with vanishing second curvature tensor II Tensor N.S. 26, (1972) pp. 250-254
[23] S. Kobayashi and K; Nomizu, Foundation of Differential Geometry, 1963, Interscience Publishers
[24] S. Kobayashi. Transformation groups in Differential Geometry, Springer – Verlag 1972
[25] N.Koiso. On the second derivation of the total scalar curvature, Osaka Journal of Math, 16 (1979) 413-421
[26] B. Kostant, Holonomy and the Lie Algebra of infinitesimal motions of a Riemann manifold, Transactions of Amer. Math. Society t. 80 , 1955 pp 528-542

[27] A. Lichnérowicz, Théorie Globale des connexions et des groupes d'Holonomie, Editioni Cremonese Roma (1954)
[28] A. Lichnérowicz. Géométrie des groupes de Transformations (Dunod Paris, 1958)
[29] M. Matsumoto. Foundations of Finsler Geometry and Special Finsler Spaces 1986. Kaiseisha Press, Japan
[30] J. Milnor. Morse Theory, Ann of Math Studies 51. (1963) Princeton University Press, Princeton.
[31] F. Moalla, Espaces de Finsler Complets (C.R. Acad. Sc. Paris, 1964, t.258, n° 8 p. 2251 et n° 10 p . 2734.
[32] Y. Muto. On Einstein Metrics. Journal of Differential Geometry 9 (1974) 521 -530
[33] M. Obata, Certain conditions for a Riemannian manifold to be isometric with a sphere. Journal Math. Society of Japan 4 (1962) 333-340
[34] B. Riemann Über die Hypothesen welche der Geometrie zugrunde liegen. Habilitationsvortrag 1854. Ges. Math. Werke, 272-287. Leipzig 1892, Reproduced by Dover Publications 1953
[35] H. Rund, The Differential Geometry of Finsler Spaces, Springer – Verlag 1959
[36] N. Steenrod, The Topology of Fibre Bundles, Princeton University Press, 1951
[37] K. Yamaguchi, On Infinitesimal Projective Transformations, Hokkaido. Math. Journal 1974 (3). P.262-270

INDEX

Printed and bound by CPI Group (UK) Ltd, Croydon, CR0 4YY

13/05/2025

01870634-0001